Fuels and Chemicals from Oilseeds

Technology and Policy Options

AAAS Selected Symposia Series

Published by Westview Press, Inc.
5500 Central Avenue, Boulder, Colorado

for the

American Association for the Advancement of Science
1776 Massachusetts Ave., N.W., Washington, D.C.

Fuels and Chemicals from Oilseeds

Technology and Policy Options

Edited by Eugene B. Shultz, Jr., and Robert P. Morgan

Routledge
Taylor & Francis Group

LONDON AND NEW YORK

First published 1984 by Westview Press

Published 2018 by Routledge
52 Vanderbilt Avenue, New York, NY 10017
2 Park Square, Milton Park, Abingdon, Oxon OX14 4RN

*Routledge is an imprint of the Taylor & Francis Group, an
informa business*

Library of Congress Catalog Card Number 83-51194

ISBN 13: 978-0-367-01491-9 (hbk)

ISBN 13: 978-0-367-16478-2 (pbk)

About the Book

With the sharp rise in petroleum prices over the past decade, interest has grown in the possible use of plant and tree seed oils as diesel fuel substitutes or extenders and as chemical feedstocks in both the industrialized and the developing world. This book represents the first attempt to bring together science, technology, and policy aspects of this topic in one volume. The authors examine the chemical, biological, and energy-related resources available in commercial and wild plant oilseeds; the status of chemical processing efforts and of engine tests using agricultural diesel fuels; the potential economic, environmental, and social impacts of more extensive oilseed utilization; and current legislation aimed at stimulating commercialization of strategic renewable agricultural resources.

About the Series

The *AAAS Selected Symposia Series* was begun in 1977 to provide a means for more permanently recording and more widely disseminating some of the valuable material which is discussed at the AAAS Annual National Meetings. The volumes in this *Series* are based on symposia held at the Meetings which address topics of current and continuing significance, both within and among the sciences, and in the areas in which science and technology impact on public policy. The *Series* format is designed to provide for rapid dissemination of information, so the papers are not typeset but are reproduced directly from the camera-copy submitted by the authors. The papers are organized and edited by the symposium arrangers who then become the editors of the various volumes. Most papers published in this *Series* are original contributions which have not been previously published, although in some cases additional papers from other sources have been added by an editor to provide a more comprehensive view of a particular topic. Symposia may be reports of new research or reviews of established work, particularly work of an interdisciplinary nature, since the AAAS Annual Meetings typically embrace the full range of the sciences and their societal implications.

WILLIAM D. CAREY
Executive Officer
American Association for
the Advancement of Science

Contents

About the Editors and Authors

Eugene B. Shultz, Jr., *is professor of engineering and applied science and associate director of the Center for Development Technology at Washington University in St. Louis, Missouri. He is active in interdisciplinary technical and economic feasibility studies of innovations in renewable sources of energy and chemicals and in technology for international development. His background is in chemical engineering and applied chemistry.*

Robert P. Morgan, *currently science policy fellow at the Brookings Institution in Washington, D.C., is professor and chairman of the Department of Technology and Human Affairs at Washington University in St. Louis, Missouri. Formally trained in chemical and nuclear engineering, he now does research on social applications and assessment of technology, science and technology for international development, and renewable energy sources. Among his publications are* Science and Technology for Development: The Role of U.S. Universities *and* Renewable Resource Utilization for Development *(Pergamon, 1979 and 1981, respectively).*

George E. Brown, Jr., *an industrial physicist by training, is U.S. Representative to Congress from Riverside, California. His primary interests are in science policy and agricultural research, and he is chairman of the Subcommittee on Department Operations, Research, and Foreign Agriculture, and a member of the Science and Technology Committee; the Subcommittee on Science, Research, and Technology; the Subcommittee on Space Science and Applications; and the Congressional Technology Assessment Board.*

Joe R. Cowles *is professor of biology at the University of Houston. His research interests are symbiosis as related to nitrogen fixation, aromatic and protein biosynthesis in plant tissues, and short-rotation silviculture systems.*

William P. Darby, *a civil engineer, is associate professor of technology and human affairs at Washington University in St. Louis, Missouri. His research interests include environmental problems associated with air, land, and water pollution and the control of environmental carcinogens; cost-benefit and risk-benefit analyses; technology assessment; and quantitative methods in government regulation and decision-making.*

Harold M. Draper III *recently earned his doctoral degree from the Department of Technology and Human Affairs, School of Engineering and Applied Science at Washington University in St. Louis, where he was also associated with the Center for Development Technology. A specialist in biomass energy technology and policy, he has published on oilseeds as energy sources in rural and developing areas.*

Cady R. Engler *is an assistant research engineer at the Food Protein Research and Development Center at Texas A&M University. He has done work on the extraction of oils and hydrocarbons from biomass and on the processing and utilization of biomass as a source of chemicals and energy.*

Kenton R. Kaufman *is assistant professor of agricultural engineering at North Dakota State University, Fargo. His interests are ergonomics and energy for alternate systems and methods of agricultural production. Currently he is doing research on the use of vegetable oils as diesel fuels.*

Robert Kleiman, *research leader at the USDA Northern Regional Research Center in Peoria, Illinois, has been working in the field of seed composition since 1962. He has discovered numerous new lipid compounds and published more than 70 papers describing his materials and instrumental methods. He is coeditor of* Recent Advances in Phytochemistry: The Resource Potential in Phytochemistry *(Plenum, 1980).*

Stephen Kresovich, *assistant professor of soil and crop sciences at Texas A&M University, has been doing research on fuels from new crops and on the physiology and breeding of seed sorghum and sugar cane.*

Edward S. Lipinsky *is senior research leader of biomass resources in the Department of Applied and Technical Economics at Battelle Columbus Laboratories in Columbus, Ohio. A specialist in biomass resources, he has directed systems studies of fuels and chemical feedstocks from sugar crops and corn; evaluation studies of biomass systems for electric generation; developmental studies of new fermentation systems using sugar* *crop and starch crop raw materials; and feasibility studies of ethanol production.*

William Lockeretz, *a specialist in agricultural resources, is a research associate in the School of Nutrition at Tufts University. He has studied energy consumption in agriculture; use of crop residues for energy; alcohol fuel production from agricultural materials; and low-energy alternatives to chemical pest control and fertilization methods. He has edited* **Agriculture and Energy** *(Academic, 1977),* **Agriculture as a Producer and Consumer of Energy** *(AAAS Selected Symposium 78; Westview, 1982), and* **Environmentally Sound Agriculture** *(Praeger, in press).*

Sandra Lee Mathieu *earned an advanced degree at the Department of Technology and Human Affairs, School of Engineering and Applied Science, Washington University in St. Louis, where she was also associated with the Center for Development Technology. Her area of specialization is utilization of biomass, especially oilseeds, as novel energy sources in developing countries. Currently she works in Houston as a consultant on the use of computers in management of biological databases.*

Thomas A. McClure *is a senior agricultural economist in the Department of Technical and Applied Economics at Battelle Columbus Laboratories in Columbus, Ohio. A specialist in agricultural finance and marketing, he is coeditor of the* **Handbook of Biosolar Resources, Vol. II: Resource Materials** *(CRC, 1981).*

James L. Otis *is a chemical engineer in the Chemistry Department's Environmental Technology Section at Battelle Columbus Laboratories in Columbus, Ohio. He carries out economic analyses and cost estimates of chemical, energy, and environmental processes and businesses.*

L. H. Princen *is associate director of the USDA Northern Regional Research Center in Peoria, Illinois. His interests are development of new crops and industrial applications of agricultural commodities and products. Currently he is president of the Illinois State Academy of Science and councilor of the American Chemical Society.*

Everett H. Pryde, *research leader of the Oilseed Crops Laboratory at the USDA Northern Regional Research Center in Peoria, Illinois, has specialized in the chemistry of fats and oils. He has more than 120 publications and patents in his field, among them* **Handbook of Soy Oil Processing and Utiliza-**

tion *(American Oil Chemists Society/American Soybean Association, 1980) and* Fatty Acids *and* New Sources of Fats and Oils *(American Oil Chemists Society, 1979 and 1981, respectively).*

Peter H. Raven, *a specialist in plant systematics and evolution, is director of the Missouri Botanical Garden, Engelmann Professor of Botany at Washington University, adjunct professor of biology at St. Louis University, adjunct professor of biology at the University of Missouri, St. Louis, and a member of the National Academy of Sciences. He has written or edited more than 200 publications in his field, including* Coevolution of Animals and Plants *(with L. Gilbert; University of Texas Press, 1975),* Research Priorities in Tropical Biology *(National Academy of Sciences, 1980), and* Biology of Plants, *3rd ed. (with R. Evert and H. Curtis; Worth, 1981).*

H. W. Scheld *is senior scientist in the Department of Biology at the University of Houston. A specialist in forestry and plant physiology, he is studying the flowering and reproductive strategy of the Chinese Tallow tree and its uses in a short-rotation intensive silviculture system, as a perennial oilseed crop, and as bee forage and a honey crop.*

Cynthia K. Wagner, *a doctoral student in the Department of Social Systems Sciences at the Wharton School, University of Pennsylvania, has specialized in technology in agriculture. She has published on fuels and organic chemicals from biomass, the potential of emergent aquatic plants as biomass for fuel and chemicals, and corn-derived fuel and animal feed production.*

Anna-Maria V. Watowich, *a chemical engineer with an advanced degree in technology and human affairs from Washington University, is a process engineer at Dow Chemical Company in Freeport, Texas. Her interests are alternative sources for chemical feedstocks, with emphasis on oilseed-derived chemicals.*

Acknowledgments

We are grateful to the many contributors to this volume for their willingness to communicate their insights and find- ings and for their assistance in preparation of the final manuscript. Our special thanks go to several members of the scientific staff of the Northern Regional Research Center, U.S. Department of Agriculture, Peoria, for their advice and counsel, based on their extensive experience with oilseeds.

We also wish to thank our fine support staff at Washington University, and especially Donna Williams for all of her excellent work; Murray Sacks assisted greatly in preparation of the Index. Our thanks also to the AAAS and Westview Press for their effort. One of us (RPM) was on leave at and aided by the Brookings Institution during the final months of book preparation.

Finally, while acknowledging our debt to others, responsibility for the findings and conclusions within specific chapters rests solely with the chapter authors.

Eugene B. Shultz, Jr.
Robert P. Morgan
St. Louis, Missouri

_____ *Eugene B. Shultz, Jr., Robert P. Morgan*

Introduction

This volume is based on the Symposium "Fuels and
Chemicals from Oilseeds: Technology and Policy Options,"
held at the Annual Meeting of the American Society for the
Advancement of Science in Washington, D.C. on January 5,
1982. That session attracted considerable attention and
received coverage in such media as Science News, Chemical
Week, and National Public Radio's program, All Things
Considered. Although perhaps not as newsworthy at the
Annual Meeting as "creation science" and genetic engineering, the topic of fuels and chemicals from oilseeds appeals
to a broad scientific audience because of its timeliness
and novelty, and because it cuts across a large number of
scientific disciplines, including agronomy, genetics,
chemistry, agricultural engineering, chemical engineering,
and economics. In addition, it raises a number of important questions concerning resource utilization, environmental quality, industrial innovation, and the role of
government.

At the time of our article on this topic in the
September 7, 1981 issue of Chemical and Engineering News,
relatively little had been written on novel oilseeds which
reached out beyond the purely technical dimension. There
was the pioneering work of researchers at the U.S. Department of Agriculture's Northern Regional Research Center
(NRRC) in Peoria, Illinois, who are represented in this
volume. In 1981, the American Oil Chemists' Society published a volume entitled New Sources of Fats and Oils,
edited by E. H. Pryde, L. H. Princen, and K. D. Mukherjee,
which is a valuable source of technical information. Subsequent to the AAAS Symposium and two earlier meetings on
Vegetable Oils as Diesel Fuel held at USDA's NRRC, an
International Conference on Plant and Vegetable Oils as
Fuels was held in August, 1982, under the aegis of the
American Society of Agricultural Engineers. Interest in

the subject of fuels and chemicals from oilseeds has continued to grow.

The present volume attempts to integrate the scientific
and technical with the policy dimension. The latter has
come to the fore because of several developments during the
last five years: the continuing concern over reliable,
accessible sources of energy and chemical feedstocks;
actions by oilseed farmers during diesel fuel shortages in
the late 1970s when some tried vegetable oil as tractor
fuel; proposed legislation in the U.S. Congress to establish an Arid Lands Renewable Agricultural Resources Corporation which would support, among other things, novel oilseed
developments in the Southwest.

In most cases, seed oils (vegetable oils) are mixtures
of triglycerides, which are esters of glycerol and fatty
acids. The term "oil" is ambiguous and has several legitimate and distinct meanings. Two that are often confused
are hydrocarbon oil, and triglyceride oil. The latter, of
vegetable or seed origin, is not a hydrocarbon -- that is,
a compound of only hydrogen and carbon -- but a mixture of
partially oxygenated hydrocarbons. An example of a petroleum-based hydrocarbon oil is diesel fuel. In this book,
"seed oil" or "vegetable oil" will refer to oils obtained
from seeds of living plant species. These are usually
triglyceride oils.

Triglyceride oils can be obtained from the seeds of
common oilseed plants such as soy, peanut, sunflower,
cotton, and castor, as well as thousands of lesser-known
oilseed plants and trees. Some triglyceride oils are
edible; others are inedible, but contain chemical functiona-
lities of industrial value such as acetylenic, epoxy, and
allenic groups, and conjugated unsaturation, in long- and
short-chain fatty acids. Seed oils may be utilized as
found, or in the form of their fatty acids. Alternatively,
seed oils may be converted to aromatic compounds -- such as
benzene, toluene, and xylenes for use as chemical raw
materials or in high-octane gasoline -- as well as to two-
to four-carbon paraffins and olefins for use as liquefied
petroleum gas or in making chemicals.

Oilseeds could conceivably be useful in replacing
firewood or charcoal used to cook food in rural areas of
developing countries experiencing deforestation problems.
Also, it is possible that oilseeds might be crushed and
refined locally, using simple, low-cost technology, and the
extracted oil burned as a kerosene substitute for lighting.
Other potential uses for seed oils as fuels include use as

a turbine fuel for peak-load generation of electricity, as fuel for home heating, and as diesel fuel.

Three or four years ago, spurred on by rising petroleum prices and local shortages of diesel fuel for farm use, interest developed among sunflower farmers as well as others in utilizing vegetable oils as substitutes or extenders for diesel fuel. Although short-term engine tests under favorable conditions were encouraging, long-term engine tests now in progress in several laboratories indicate that much work remains to be performed before vegetable oils or their derivatives, such as esters, can be accepted as satisfactory diesel fuels. Vegetable oils, at the least, must be well-filtered and freed of gums before use. Existing diesel engines with precombustion chambers perform better than direct-injection diesel engines commonly used in agriculture, in part because carbon deposition due to the high viscosity of seed oils is more severe in engines of the direct-injection type. Further, unsaturated oils that pass the piston rings and enter the crankcase may polymerize. In short, a number of technical problems can plague those who may take a casual approach to the use of seed oils as diesel fuels, even if the seed oil is mixed with a generous quantity of diesel oil. South Africa and Brazil are already well along in learning how to utilize vegetable oils in diesel equipment and, coupled with the engine testing now going on in U.S. university, government, and industrial laboratories, the likelihood is growing that the technical problems will not prove insurmountable for vegetable oils to be used as diesel fuel extenders, if not as substitutes. However, the recent leveling and falling off of petroleum prices and its relatively abundant short-term supply cast doubt upon whether seed oils will become economically feasible as diesel fuel extenders in the near future.

In this volume, the six original symposium papers, modified and updated, have been combined with additional papers to form a comprehensive and timely treatment of a fascinating and important topic. In Chapter One, Congressman George Brown of California, the member of Congress most knowledgable about and involved in science, technology and agriculture activities in the Congress, presents the case for utilizing novel crops on arid lands, including oilseeds, and discusses legislation introduced to facilitate such utilization. In Chapter 2, we and our colleagues, William Darby and Harold Draper, assess the potential for bringing about widespread utilization of novel oilseeds on marginal lands. The benefits to be derived from such a scenario as well as the risks and obstacles are included. In Chapter 3, Peter Raven, Director of the Missouri Botanical Garden,

stresses the importance of research to learn about the many
underexplored plant varieties in the world before some are
irrevocably lost to modernization.

In Chapters 4 and 5, scientists at the USDA Northern
Regional Research Center summarize developments in two areas.
First, Everett Pryde discusses chemicals and fuels from
commercial oilseed crops; then Robert Kleiman and L. H.
Princen of the USDA are joined by Harold Draper of Washington
University in examining the potential for chemicals and
fuels from wild plant oilseeds, including some of possible
significance in tropical countries. In Chapter 6, William
Scheld and co-authors present a wealth of information on
one novel oilseed, the Chinese tallow tree which, because
of its potentially high yields of two different fat and oil
materials, is currently under intensive study as a commercial oilseed tree crop of the future. Anna-Maria Watowich
and E. B. Shultz, Jr. examine certain aspects of chemical
feedstock uses of seed oils in Chapter 7, focusing on the
market potential for commercialization of the higher nylons
and related products from novel oilseeds that might be grown
in the southeastern United States.

Spurred on by the interest in utilization of sunflower
oil as a diesel fuel substitute or extender by farmers, the
U.S. Department of Agriculture has funded a variety of
technical studies. Kenton Kaufman evaluates sunflower oil
as a diesel fuel in Chapter 8 and describes and evaluates
diesel engine testing experience with vegetable oils in
Chapter 9. Economic aspects of oilseed utilization for
diesel fuel, involving either centralized or on-farm extraction, are analyzed by William Lockeretz in Chapter 10.
E. S. Lipinsky and colleagues, in Chapter 11, assess the
production potential for fuels from oilseeds in the U.S.,
using a variety of cropping schemes.

The potential for and some initial experiments on the
use of oilseeds as cooking and lighting fuel in developing
countries where firewood is scarce and kerosene is expensive
are examined in Chapter 12 by Sandra Mathieu and E. B.
Shultz, Jr. Finally, we have added some concluding remarks
to summarize the state-of-the-art and bring closure to the
volume.

1. Renewable Resources: Opportunities and Legislative Initiatives

Introduction

Over the years, I have experienced much frustration trying to interest people in underutilized crops such as the novel oilseed crops being discussed in this volume. Perhaps as a result of sessions such as the one on oilseeds at the AAAS Meeting in January, 1982, and the increased publicity that this subject is receiving, we will begin to see some of these unconventional ideas woven into traditional agricultural and industrial practices.

It seems strange to me that this blossoming of interest has not yet fully occurred. These novel crops and products seem to meet a number of pressing national needs. They can provide cropping alternatives to farmers and alternative sources of feedstocks to industry. In many cases, they can be raised on marginal lands or can form the basis of a more adaptive style of agriculture that is less resource intensive. They can provide a secure, domestic source of essential and strategic materials. And, they are a source of replacements for both fuel and non-fuel end uses of petroleum.

I have tried, on a number of occasions, to weave some of these ideas into legislative proposals and have met with mixed success. I will frankly admit that it is probably too early for some of these ideas to take hold within industry or even within the traditional research system. But given the long lead time required to make changes in agricultural practices, we must begin today to research and develop alternatives such as these being discussed.

The Need for Renewable Resources
and Oilseeds

During times of change, mature bureaucratic systems tend to become reactionary, in that they tend to seek solutions to present problems by using traditional approaches. In the category of mature systems exhibiting reactionary behavior I include Congress, government, and established industries. These systems only develop new responses to change well after the point at which a dispassionate observer or an adequate job of planning would indicate that new approaches were required.

A good example of this is our current approach to dealing with petroleum cost and supply problems. Instead of heavily emphasizing conservation and the development of renewable sources to replace petroleum end uses, Congress has placed a heavy emphasis on the synthetic fuels program and sought other new sources of crude oil.

Another example is our current "manipulative" style of agriculture, heavily dependent upon non-renewable resource inputs. We have had great success in manipulating our crop ecosystem with investments of energy, chemicals, and water. Too much success, apparently, because in spite of the fact that these inputs are becoming increasingly expensive and scarce, our agricultural policies continue to encourage traditional cropping systems and the planting of traditional crops.

The novel oilseed crops and their products being discussed are examples of innovative contributions toward dealing with both of these major problems. It is in the national interest that research and development work in this area continue and be expanded.

For the farmer, these plants can provide needed cropping alternatives. Our agricultural system is heavily dependent upon a limited number of crops. Many areas of our country practice monoculture and their economies are extremely vulnerable to weather and market disruptions. The development of new crops, such as these novel oilseed crops, and the development of new markets for existing oilseed crops would greatly benefit the farmer. Differences in harvest schedules, storage requirements, and markets between some of these novel crops and the traditional crops could allow farmers valuable alternatives. The industrial sector would benefit as well, with new industries developing around these crops when they become commercialized.

Many of these crops grow on marginal land and require a
minimum of inputs. Many of them are perennials and require
less intensive cultivation than annuals. The commercial
development of these crops would provide farmers with a form
of crop insurance. Bad years for traditional crops may not
as seriously affect these novel crops and might help the
farmer survive an otherwise disastrous year. Many of these
crops can replace resource depleting traditional crops being
grown on marginal lands. There is the potential as well for
using these crops to bring new acreage into production which
is unable to support traditional crops.

This last aspect of many of these novel crops brings up
the potential of changing our current farming practices. As
I stated earlier, our current farming practices are based
upon the use of limited resources to manipulate the crop
ecosystem. With some of these novel crops, the potential
exists to develop adaptive farming systems, where the crop
to be planted is selected according to its suitability for
the existing agroclimatic conditions. Instead of producing
cotton on marginal croplands in the arid Southwest, we could
produce jojoba or buffalo gourd at a fraction of the irrigation demand, conserving declining water resources, while
continuing to provide a source of income for the farmer.
Also, many of these native crops are halophytic (salt tolerant) and could provide on-farm alternatives to the various
capital-intensive desalinization schemes being discussed in
the Southwest. A little imagination brings a wealth of
these adaptive approaches to light.

These novel oilseeds are of national importance if we
are serious about developing domestic sources of critical
industrial and strategic materials. Many of these oilseed
crops produce substances essential for domestic industries.
Many of these substances are of such importance that they
are included in the national strategic stockpile. Domestic
sources of these substances can reduce the need to actually
have them in storage and can provide a reliable supply of
these feedstocks for industry.

Of course, the major importance of these oilseed crops
lies in their potential for displacing petrochemical-based
materials. Many of these oilseed crops, both novel as well
as traditional crops, such as sunflower, peanut, soybean,
and other oils might become practical replacements or extenders of diesel fuel. But there are a number of other
applications which could replace non-fuel petroleum use.
Chemical feedstocks, lubricants, plasticizers, and other
essential industrial materials can be produced from domestic

oilseeds to replace petrochemicals. These non-fuel uses of
domestic oilseed production are very important because in
considering these alternatives we begin to focus on replacing petroleum end uses and get away from our myopic emphasis
on simply producing more crude oil. By starting upstream in
our petroleum and petrochemical distribution system and working back, a variety of alternatives become available to us.

Legislation

But, as I stated earlier, the rationality of these reasons to actively pursue research and development work on
these approaches is lost on the mature institutions which
set national policy. In the 1977 Farm Bill, I inserted
language which directed the Department of Agriculture to
begin research on new crops such as jojoba and guayule.
Some funds were devoted to this work, but not nearly enough
to achieve any real progress. There was simply too much
competition for these research and development funds from
conventional crop programs.

In 1978, I authored and saw enacted the Native Latex
Commercialization and Economic Development Act of 1978.
This law established a research and demonstration program
for guayule, a semiarid shrub which produces natural rubber.
I had hoped that the success of this work on a specific
novel crop would open the door for serious consideration of
the potential of other underutilized crops, such as oilseeds.
However, again this new effort did not have the constituency
required to succeed in budget battles with traditional agricultural research and few gains have been made.

In the 1981-1982 Farm Bill, a new category of competitive research grants was set up within USDA to study new
crops, such as jojoba and guayule. This program will be
established when funds above the current levels are attained,
admittedly a difficult task. But the authorization is there
and there is some support on the House Appropriations Subcommittee for research on some of these new crops, especially
if the benefits to the farmer are made apparent. If enough
enthusiasm for this competitive research grants program can
be generated, there is now the possibility of research funds
being dedicated to the crops we are discussing in this symposium.

The major focus of efforts to speed the development of
underutilized crops at present is the bill to establish an
Arid Lands Renewable Agricultural Resources Corporation.
This legislation which Senator DeConcini and I have introduced in our respective bodies of Congress, would establish

a corporation identical to the Synthetic Fuels Corporation but one dedicated to promoting crops native or adaptable to semi-arid regions. Included among these are jojoba and buffalo gourd, both promising oilseed crops.

Senator DeConcini and I feel that there is an excessive emphasis in our national programs on humid-region sources of petroleum replacements. We also feel that an inordinate amount of attention is being directed toward finding crude oil replacements and inadequate attention is being devoted toward replacing petroleum end uses with renewable sources. The synfuels projects scheduled for the arid West are completely inappropriate for our region and we feel that alternatives must be presented.

Again, the intent of this legislation is to show Congress and the federal government that the development of these novel crops can greatly enhance the national effort to free us from our dependence upon imported oil. We know that there is a great deal of commercial interest in these innovative approaches but the returns are too far in the future for any corporation to justify great R&D expenditures in this field. With a minimal amount of government expense, using this Arid Lands Corporation as one model, we can bridge existing research into the commercial sector.

From Research to Commercialization

It is essential that industry, universities, and government work together in efforts to develop these novel oilseed crops and to develop new uses for existing oilseed products. The transition from research to commercialization is the most difficult step to achieve in the introduction of any new product or technology. With other new crops, there is typically much commercial interest in the post-harvest processing and development and marketing of the products derived from these crops, but little interest in doing the agronomic research, where proprietary rights and economic returns are less certain.

This creates a difficult gap to bridge since no company will switch to a new source of feedstocks unless a steady supply can be assured. And without a guaranteed market, farmers will not raise novel crops or dedicate a portion of their harvest to some new market. If these novel oilseed crops are to become commercialized, researchers in processing and product development must work together with agricultural researchers. And the federal government must be willing to take an active role in this commercialization process.

The federal role in such an effort need not involve extensive funds, a key point in this time of fiscal restraint. What it would involve is realization of the importance of a commercialization program to develop these oilseed crops and the secure, renewable, domestic source of materials that these crops can provide. An emphasis on the national security aspects of this effort would be another key point in the present political climate. Sessions such as the AAAS Oilseed Symposium, expanded to include more of the manufacturing industry and the farm sector would be helpful in making these points.

Once sufficient interest were generated, federal agencies could form a commercialization task force to coordinate their R&D funds on this area. Such an effort could be put together by the Office of Science and Technology Policy. Or, if specific authorization were needed, an entity similar to the Joint Commission on Native Latex could be established. While the latter has its flaws, it has managed to provide a clearinghouse for information and a focal point for government-wide efforts to commercialize guayule. And it also serves as the focus for government-industry cooperation.

Such a federal effort could provide the "glue money" for the industry-university-government cooperation in a generic research and development program on novel oilseeds and novel oilseed uses. The federal share of this R&D work could be current federal research programs and additional, modestly funded programs such as the recently authorized competitive research grant category on new crops set up in USDA. Through the Cooperative State Research Service in USDA, federal and state funds could be combined at land-grant institutions to perform additional work.

Other areas for federal initiatives could involve our foreign aid programs. One of the most pressing problems in developing and transitional countries is the development of renewable sources of fuel and the development of resource-conserving industries. Many of these countries desperately need a sustainable agricultural base and, where food crops could not be grown, many of these novel oilseed crops could provide that base.

Many of the plants we are discussing can be raised in these countries, in fact many are native to them. Jojoba and buffalo gourd can be raised in many arid tropical countries. Chinese tallow tree is native to the Asian humid tropics. Much crop development work can be done under our aid programs and additional novel oilseed crops could be

identified as well. Although I am touching lightly on the
international aspects, it could well be that this area holds
the greatest immediate promise for the advancement of work on
these novel crops.

Industry involvement could be in the form of closely coordinated, wholly privately funded research, or could take
the form of industry-university matching research efforts.
To insure success, a broad spectrum of industry would have
to be involved. From the farm sector, funding from soybean
oil, peanut, sunflower and other oilseed growing and processing industries should be sought. Farm cooperatives and
other farm groups with research and promotional funds should
be included. From other industrial sectors, such as the
chemical industry, the petroleum industry, and other end
users, interest already exists but needs to be increased.

In this admittedly simplistic scenario, the needed agronomic research should be closely coordinated with processing
and product development work. The results of field trials
and plant selection could be fed to processing researchers
and their findings then given back to the plant researchers
to aid them in further research modifications.

At the same time, the federal government, a major market for any materials for products developed, could be inven-
torying its procurement needs. Under the Native Latex program, there is a committee in the Defense Department com-
piling a listing of DOD natural rubber needs to see if future
DOD procurements could use guayule rubber. A similar effort
in oilseed products could show the extent of the potential
market and prepare the way for a domestic, renewable procurement program in the federal government. We already have a
"Buy American" program for federal purchases, and there is
every reason to expand that to a "Buy American Renewables"
program.

At the point when the threshold of commercial development
was reached, the Arid Lands Renewable Agricultural Resources
Corporation would play a key role. Even though the promise
of these novel crops and their products would be apparent,
and the reactionary institutions referred to earlier would
have now become merely conservative, a large gap would still
have to be bridged. Banks would not be willing to invest
funds on their own. Crop insurance and farmer operating
loans would be difficult to obtain for raising these novel
crops. Existing capital investments, standing feedstock
supply contracts, and other industry arrangements would sus-
stain inertia in the private sector, preventing a major
shift of corporate resources.

With a modest federal investment, and investments made
for a limited amount of time, this Corporation could be the
bridge to commercial development of these crops. The Corporation would use price guarantees, loan guarantees, direct
loans, or enter into joint ventures to foster the growing,
marketing, and processing of some of the crops we are discussing. And with a sunset provision and a board of directors
empowered to shift the Corporation's activities, there would
be little danger of permanent subsidization or artificial
market conditions being created.

But whatever methods or institutional structures are
used, I feel that the development of underutilized plant
resources such as these novel oilseeds is extremely important.
In detailing what I envision as one way to proceed toward
commercialization, I am not excluding other approaches. I
hope that one outcome of this book will be the emergence
of a number of ways to achieve the development of new
crops and new crop uses. I offer my assistance in Congress
for any reasonable program toward this end. After all, I
would not want it said of me that I am a "reactionary"
thinker.

*Eugene B. Shultz, Jr., William P. Darby,
Harold M. Draper III, Robert P. Morgan*

2. Novel Marginal-Land Oilseeds: Potential Benefits and Risks

Introduction

Oilseed plants and trees, cultivated and wild, generally exhibit a remarkable diversity of characteristics which make them potentially attractive for fuel and chemical uses. However, most of the thousands of plant species that bear seeds with attractively high oil contents have been given little development attention. Many oilseeds are found in temperate or tropical countries, thriving on otherwise unproductive marginal lands in dry, wet, hilly, nutrient-poor or saline soils. Therefore, some of these novel marginal-land oilseed plants and trees deserve attention for enlargement of the global agricultural base.

If such oilseeds become domesticated and widely adopted for chemical and fuel farming on marginal lands, some of the direct pressure on good lands needed for food production might be avoided. However, where fragile marginal-land ecosystems are to be opened to agriculture, careful planning and management must be provided to prevent soil damage and productivity losses which in many cases could be irreversible. Fertility of marginal land often declines rapidly with use, depending on many factors including the willingness and ability of the farmer to use the best available cropping practices. As scientific data on the genetics, agronomy and utilization of these novel plants and trees are being accumulated, and cost-effectiveness evaluations are being carried out, we feel that social, economic and ecological consequences should be assessed concurrently. Novel oilseed developments for marginal lands are still in early stages of study, so that important indirect, unanticipated or delayed consequences may still be identified, and appropriate corrective measures developed and applied in time.

Whether marginal-land oilseeds can contribute in a
major way to future fuel and chemical stocks with ecological
as well as economic acceptability is an important issue for
study, with many ramifications. Adoption of marginal-land
oilseeds could lead to the cultivation of millions of presently idle acres, greatly increasing the demand for water,
fertilizer, pesticides and other inputs. Many industries
would be affected significantly, such as farm equipment,
fertilizer, and pesticides. In the United States, there
would be less dependence on imported oil and more reliance
on renewable resources under domestic control, but more
ecologically-fragile land would be in jeopardy. Marginal-
land agriculture if carried out correctly may bring substantial benefits to the soil, as well as direct economic
benefits to marginal-land farmers. It is possible in many
cases in the Third World that processes of desertification
and deforestation might be slowed. Oilseed perennials and
trees would generally cause little soil disturbance, compared to cropping of annuals. If vegetative residues could
be returned to the soil, soil tilth and fertility might be
improved.

Marginal-land oilseeds should not be regarded as low-
yielding oilseeds because they grow on less than prime
farmland. As is shown in Table 1, buffalo gourd may produce
about as much oil on dry land as peanut and safflower can
yield on good land, and Chinese tallow tree may produce much
more oil per acre on wet lands than any conventional tem-
perate-zone oilseed can produce on good farmland.

Marginal-Land Oilseeds for
the Southwestern United States

In this paper, the Southwest is considered to be
western Texas, the Oklahoma panhandle, southwest Kansas,
southwest and southeast Colorado, New Mexico, Arizona,
southern Nevada, southern Utah, and southern California.

As a specific example of a marginal-land oilseed under
development for commercialization, consider the buffalo
gourd (Cucurbita foetidissima) which is potentially cultivatable on dry lands such as are found in Arizona, New
Mexico, west Texas, Mexico, and some Third World dry lands.
This presently uncultivated plant indigenous to the southwest United States and northern Mexico (Figure 1) has the
remarkable ability to withstand hot dry growing conditions
and still produce oil yields per acre that are potentially
greater than those obtained from sunflower growing in North
Dakota and Minnesota on better soil with much more water.
The oil, after refining, may be suitable for diesel fuel or

Table 1. Approximate Yields from Typical Oilseeds

	Annual yield in the U.S., barrels per acre
[a]Chinese tallow tree	12[b]
[a]Buffalo gourd	2[c]
Peanut	2
Safflower	2
Rapeseed	1
Sunflower	1
Soybean	0.9

Source: Reference 1.

[a]Marginal-land oilseeds.. Others are conventional oilseeds requiring good land.

[b]Includes the solid fat on the outside of the seed, as well as the kernel oil. Potential yield based on estimates from wild stands (2).

[c]Estimated potential yield (3).

for food or chemical purposes, and the co-product seed meal is high in protein. In addition, buffalo gourd develops an extensive starchy root system that would probably have to be harvested periodically to prevent overcrowding of the field. We estimate that a potential 400 gallons of alcohol per acre (34 million Btu per acre) might be obtained by conventional fermentation of the root starch (4). Alternatively, the starch might be used by the food industry. The buffalo gourd is under development at the University of Arizona, by W. P. Bemis and his colleagues, a group with several years of hybridization experience with this plant (5).

The considerations presented above suggest that buffalo gourd might become an important commercial oil, starch and protein crop for hot dry lands. Buffalo gourd will likely require only a fraction of the irrigation water that such crops as corn and wheat need, and therefore might replace those and other water demanding crops in areas where there is danger of exhaustion of natural aquifers, for example, in the Texas High Plains (3).

Bemis and co-workers have proposed a novel cropping system to optimize root starch yields and prevent overcrowding

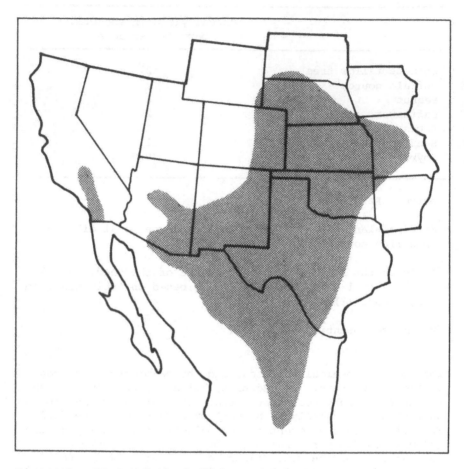

Figure 1.   Range of the buffalo gourd in the southwestern
United States (courtesy of William P. Bemis, University of
Arizona).

of the plantation (5b). Buffalo gourd develops extensive
systems of large fleshy roots as means of adapting to hot
dry conditions. Weights of over 100 pounds per root can be
attained in just three or four seasons. Because of repro-
duction by adventitious rooting from the extensive and
numerous runners, development of roots could severely crowd
the plantation if roots are not harvested periodically. The
proposed system includes the digging of alternate one-meter-
wide swaths annually, to remove all roots in the swath. The
next year, the harvested swath is regenerated by asexual
rooting of the vines that creep over from neighboring unhar-
vested swaths. Although root harvesting in the fall will
disturb the soil, raising the possibility of wind erosion of
the soil during the winter fallow, we think it should be
possible to use vine residues to cover the swaths that have
been dug so that the effects of sustained high winds in such
places as the Texas High Plains might be mitigated.

The buffalo gourd is one of four arid-land plants that
would receive substantial government support under legislation such as the proposed Arid Lands Renewable Agricultural
Resources Corporation Act of 1981, sponsored in the House by
Congressmen George E. Brown, Jr. (D-Calif.) and Kika de la
Garza (D-Tex.). The Senate version was sponsored by Senators
Dennis DeConcini and Barry Goldwater of Arizona (6). If
such legislation became law, price guarantees, loan guarantees, direct loans and joint ventures might be made available
to encourage the production of buffalo gourd along with
guayule, jojoba, and Euphorbia lathyris. The object would
be to secure the domestic availability of these renewable
resources of fuels and chemical feedstocks, in the national
interest. Additional information is provided by Congressman
Brown in Chapter 1.

Although this proposed legislation is presently limited
to four specific arid-land plants, we recommend that it be
broadened in scope somewhat to permit other promising dry-
land oilseeds to be studied. For example, there are many
other oilseeds that grow in the Southwest, some of which
might be attractive for crop development (Table 2). Of
interest are hot-dry-land cucurbits related to buffalo
gourd, e.g., Cucurbita digitata, Cucurbita palmata, and
Apodanthera undulata which provide inedible seed oils with
conjugated triene function (7,8). Drying oils and chemical
intermediates might be obtained from these plants.

Also of considerable interest for arid lands are
drought-tolerant Lesquerella species. Lesquerella is found
in many of the same counties in the southwest U.S. as buffalo
gourd, but usually at higher elevations (9). A wild stand

Table 2.   Potential New Oilseed Crops for the Southwestern United States

A. Conjugated unsaturated acid crops

Apodanthera undulata (Melon Loco).   Seeds 31% oil, 12.7%-30% of which is a conjugated triene
and about 5% of which is a conjugated diene.   Such compounds are of interest as intermediates
for novel polymer production.   A perennial xerophytic cucurbit with a large thick root
that grows to 1 meter in length and contains 23% starch.   Habitat is dry plains and mesas
of southern Arizona, New Mexico, southwest Texas, and adjacent Mexico.

Chilopsis linearis (Desert Willow).   Seed kernels contain 31-35% oil, and oil contains a conjugated triene.   A tree that grows in sandy washes and near springs in southern Nevada,
southern California, and Texas.

Cucurbita digitata.   Seed kernels contain 28% oil, 17% of which is conjugated triene.   A perennial
xerophytic cucurbit which grows in sandy washes and loose gravel soils in northern Sonora,
southern Arizona, New Mexico, and at higher elevations in northern Baja California.

Cucurbita palmata.   Seeds contain 27-32% oil, 16-26% of which is conjugated triene and and 10% of
which is conjugated diene.   A perennial xerophytic cucurbit that grows in sandy washes and
loose gravel soils in northeastern Baja California, southern California desert north to the
San Joaquin Valley, and southwestern Arizona.

Ibervillea sonorae.   Seeds contain 28.5% oil.   Oil contains a conjugated triene.   A xerophytic
perennial cucurbit that grows on sandy, gravelly, or rocky soil in Baja California, islands
in the Gulf of California, and on plains in the thorn forest in southern Sonora.

Table 2, continued

B. Dimorphecolic acid (conjugated hydroxy diene) crops

Osteospermum ecklonis (African Daisy). Seeds 48-50% oil that contains dimorphecolic acid, a possible precursor of novel polymers. Seed yields can approach 1350 lb/acre. Planted along highways in southern California as a ground cover.

Osteospermum fruiticosum (African Daisy). Seed 13% oil, 61% of which is dimorphecolic acid. A perennial herb or shrub that is also planted along highways in California.

C. Hydroxy acid crops

Lesquerella fendleri (Yellow Top). Seeds 20-28% oil, 57-62% of which is a C-20 hydroxy acid for production of higher nylons and other polymers. This plant is a winter annual that grows on calcareous sandy soils, often forming massive populations in southwestern Texas, but also occurring in the Texas High Plains, north to Kansas and Colorado and south to north central Mexico. Yields 1000 lb seed per acre.

Lesquerella gordonii. Seeds 26-28% oil, 61-66% of which is a C-20 hydroxy acid. A winter annual that grows on sandy, disturbed soils in central Kansas, western Oklahoma, Texas panhandle, central Arizona and northern Mexico. Single populations may cover acres of land with millions of individuals. Good productivity.

Lesquerella ovalifolia. Seeds 24% oil, 62% of which is a C-20 hydroxy acid. A perennial that flowers in April, common on rocky knolls and calcareous areas in northeastern New Mexico, southeastern Colorado, the Texas panhandle and parts of Oklahoma and Kansas. Good seed production and retention.

Table 2, continued

D. Epoxy acid and acetylenic acid crops

Crepis intermedia. Seeds 19% oil, 35% of which is epoxidized oleic acid and 10% of which is crepenynic acid, chemicals useful in the polymers and protective coatings industries. This is a perennial with a long taproot, abundant mostly in the Sierra Nevada but also into northern Arizona in the Grand Canyon area. Has good agronomic characteristics but slow rates of development.

Crepis occidentalis. Variable in oil composition, but one sample had 22% oil, 30% of which was epoxidized oleic acid and 11% of which was crepenynic acid. A perennial with a taproot that is found mostly in the southwestern mountains. Has good agronomic characteristics but slow development.

Comandra pallida (Bastard Toadflax). Seeds contain 24.5% oil, and infrared analysis showed the presence of acetylenic fatty acids and trans unsaturation. This perennial herb is parasitic on the roots of other plants but is sometimes transplanted from the wild. It grows mostly on dry slopes down to 1000 feet in the mountains of California; also in the mountains of central Arizona.

Source: Adapted from Draper (25), a comprehensive compilation of information on oilseeds of potential economic value.

of L. fendleri has been successfully harvested by combine,
suggesting the potential feasibility of cropping this plant
in dry areas (10). The seed oil of lesquerella contains a
large percentage of 20-carbon hydroxy acid which might be a
source of Nylon 13, probably via cleavage between carbons 13
and 14, analogous to the way Nylon 11 is obtained from the
shorter hydroxy acid found in castor oil. Because castor
beans and their defatted meal are both allergenic (11,12),
it is unlikely that castor will be an important oilseed crop
in many countries. Therefore, Nylon 13, which has many
properties similar to those of Nylon 11 and which might be
domestically produced from southwestern lesquerella, probably
deserves more attention. John A. Rothfus of the Northern
Regional Research Center, U.S. Department of Agriculture,
Peoria, is responsible with his colleagues for much of the
current knowledge of the preparation of Nylon 13 and 13/13,
interesting engineering thermoplastics still awaiting commercialization (13).

Buffalo gourd and some other arid-land oilseeds may
have important advantages over another arid-land plant,
Euphorbia lathyris (gopher plant) as an energy crop in the
Southwest. The total Btu yield per acre from the perennial
buffalo gourd (14) may exceed or be similar to that from the
gopher plant, an annual. Further, the entire gopher plant
will have to be harvested and transported to large central
facilities so that the plant can be macerated and solvent-
extracted to recover oil found throughout the plant. It
would probably be uneconomic to return the solid residue,
the bulk of the plant, to the fields for benefitting the
soil. By contrast, only the seeds and roots of the buffalo
gourd need be removed from the field. The extensive vine
growth can be applied to the lands immediately after harvest
to assist in the stabilization and fertilization of the
soil.

Marginal-Land Oilseeds for
the Southeastern United States

In this paper, the Southeast is considered to be
Virginia, North Carolina, South Carolina, Georgia, northern
Florida, Alabama, Mississippi, Louisiana, eastern Texas,
eastern Oklahoma, Arkansas, southern Missouri, Kentucky,
Tennessee, and West Virginia.

Another specific example of an oilseed approaching
commercialization is the Chinese tallow tree (Sapium
sebiferum), probably one of the world's highest-yielding
oilseeds, of those that have been studied. This fast-
maturing tree grows well in warm, waterlogged saline soils

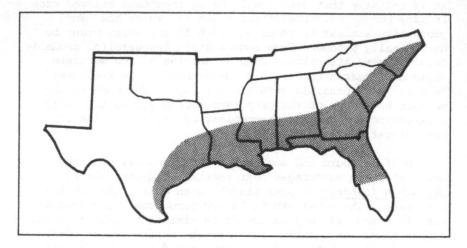

Figure 2. Approximate range of the Chinese tallow tree in
the southeastern United States (courtesy of H. William Scheld,
University of Houston).

in the southeastern United States (Figure 2) as well as in
China and India (1,15,16). The Chinese tallow tree produces
an edible solid fat on the exterior of its seeds, and an
inedible oil inside the kernel. A number of potentially
attractive industrial values are under consideration by our
group including food uses for the solid fat as a component
of cocoa butter equivalent (CBE), and chemical intermediate
uses for the kernel oil. Of particular interest for specialty polymer and pharmaceutical applications is a rare
short-chain allenic fatty acid available from the kernel
oil. Research on the domestication of the Chinese tallow
tree for seed production is being led by H. W. Scheld of the
Biology Department of the University of Houston, and Simco,
Inc. (see Chapter 6).

As in the buffalo gourd case where uncertainty sur-
rounds the impact of large-scale farming on dry-land soil
systems, there is more to be learned about the potential
problems of large-scale plantationing of Chinese tallow tree
on wet lands, including pest and disease resistance in
monoculture. In the United States where the Chinese tallow
tree is naturalized, apparently no serious pest or disease
problems have developed as yet. However, in China, the
tallow tree moth, the ailanthus silkworm, and certain aphids
are known pests (17). In the event that pests might find
their way to the U.S., control methods and their delivery
systems must be developed in advance.

Many other oilseeds may have promise for the Southeast,
especially in Appalachian, Piedmont and Coastal Plain areas
of the south Atlantic coast. Of special interest are the
possibilities of aiding the limited-resource marginal-land
farmer in these areas. Oilseeds that might be grown in
capital-conserving farming modes yielding high-valued products should be especially attractive. The introduction of
novel new high-valued oilseed crops into this region might
diversify and extend the agricultural base, as well as
benefit the limited-resource farmer. Table 3 lists potential oilseed crops for the Southeast that may be worthy of
consideration.

Table 4 provides an introduction to the potential products from marginal-land oilseeds that might be grown in
North Carolina and several other southeastern states. The
emphasis is on specialty chemicals from novel oilseeds
adapted to the particular characteristics of the soils of
the mountains, the red clay of the Piedmont, and the wetlands of the Coastal Plain which are unsuitable for large-
scale cultivation of the major food crops of the United
States. Chapter 7 deals specifically with the possibility

Table 3. Potential New Oilseed Crops for the Southeastern United States

A. Erucic acid crops

Eruca vesicaria (Eruca sativa) (Rocket Salad). Seeds 30-44% oil, 32-52% of which is erucic acid, for specialty polymers, and plasticizers. An annual plant grown as a crop in India; can be grown as a winter annual or early spring crop in the southeastern U.S. Good seed retention, good vigor, and abundant fruit set.

Iberis amara (Rocket Candytuft). Seed oil contains 38% erucic acid. A bushy annual plant that produces the large, fragrant flowers used by florists. Many cultivars, good seed retention. Can be grown as a winter annual or an early spring crop, blooming two months from germination.

Thlaspi perfoliatum (Pennycress). Seed oil contains 20-29% erucic acid and 19% C-24 monoene. A winter annual or biennial naturalized in the northern Piedmont of North Carolina. It is weak-stemmed, which may affect its crop potential.

B. Short-chain fatty acid crops

Cuphea carthagenensis (Waxweed). Seeds 32.7% oil, 62.5% of which is lauric acid, for synthetic lubricant manufacture, and for surfactants. An annual adapted to wet places in the coastal plain; habitat is marshes, ditches, and low meadows.

Cuphea procumbens (Waxweed). Seed oil contains 80% capric acid, used to make certain synthetic lubricants. An annual attributed to North Carolina.

Lindera benzoin (Spicebush). A deciduous shrub. Seeds 61% oil. High in lauric acid. Spicebush is found along streams and in alluvial woods in the Piedmont and mountains, sometimes in dense stands. Seeds can be sown in the fall, and germination is 70-80% the following spring.

Table 3, continued

C. Epoxy acid crops

Crepis vesicaria ssp. taraxifolia (Beaked Hawksbeard). Seeds 20% oil, 47% of which is epoxidized oleic acid, for the polymer and protective coatings industries. A biennial or perennial native to western Europe that produces many flowers and seeds, but the seeds shatter. It is abundant on lawns in the mountains of North Carolina.

Euphorbia lagascae. Seeds 45-48% oil, 60-65% of which is epoxy acids. An annual from Spain that grows well in North Carolina, producing seed yields as high as 950 kg/ha.

Stokesia laevis (Stokes Aster). Seeds contain 40% oil, 70% of which is epoxy acid. Seed yields as high as 1160 kg/ha. A perennial herb native to wet places in the south such as pitcher-plant bogs. Stokes aster is widely cultivated in well-drained soils that are kept moist. There are several cultivars, and seeds are available from seed companies in the United States. It has been collected wild, presumably as an escape, in Guilford Co., NC.

D. Acetylenic acid crops

Crepis alpina. Seeds contain 16-18% oil, 75% of which is crepenynic acid, potentially useful in new polymers and synthetic lubricants. This annual from Turkey has good crop potential, and grows well in South Carolina and Maryland. Seed yields have been as high as 1800 kg/ha.

Pyrularia pubera (Buffalo Nut, Oilnut). Seed oil contains both a 17 and an 18-carbon acetylenic acid. A deciduous shrub which is a root parasite on other deciduous trees and shrubs in the southern Appalachians and adjacent Piedmont. It can be cultivated like the sandalwood tree of India, which is also a root parasite.

Table 3, continued

E. *Allenic acid crops*

Sapium sebiferum (Chinese tallow tree). Nearly 50% of seed is kernel oil, plus fat on outer surface of hull. Seed yields can be as high as 5.6 tons/acre. A deciduous tree which grows profusely on saline soils and along salt marshes and shell deposits from the Texas gulf coast to North Carolina's coastal plain.

Sebastiana ligustrina. Kernel oil is similar in chemical composition to that of S. sebiferum. Grows in swamp forests and along rivers in the southeastern United States including North Carolina.

F. *Conjugated unsaturated acid crops*

Calendula officinalis (Pot Marigold). Seeds contain 43% oil, 47% of which is a conjugated triene, of potential value for novel polymers and coatings. An annual which endures frost and light snow.

Centranthus macrosiphon (Long-Spurred Valerian). 56% of seed oil is a C-18 conjugated triene. An annual from Spain which can be grown in the Southeast. Poor seed retention and very indeterminate nature, but medium productivity.

G. *Eicosenoic acid crops*

Marshallia caespitosa (Barbara's buttons). Perennial herb of Texas, Mississippi, Arkansas, Oklahoma and Louisiana. The seed oil is 44% eicosenoic acid, potentially useful as a source of higher nylons.

Table 3, continued

Lepidium virginicum (Poor-man's pepper). Winter annual. Common weed of fields, gardens and disturbed habitats of the Southeast. The seed oil is 42% eicosenoic acid.

H. Hydroxy acid crops

Lesquerella globosa. Perennial of central Tennessee and north-central Kentucky. Seeds contain
39% oil, 66% of which is probably lesquerolic acid, a potential precursor of higher nylons.

Lesquerella densipila. Annual of Tennessee and Alabama. Seeds contain 24% oil, half being
unsaturated hydroxy acids.

I. Petroselinic acid crops

Foeniculum vulgare (Fennel). Annual herb of the Southeast. Seeds contain 24% oil, 73% of which
is petroselinic acid which can be cleaved to lauric and adipic acids. The former is useful
in synthetic lubricants and surfactants; the latter in polymer manufacture.

Aralia spinosa (Hercules club). Cultivatable shrub or tree of the Southeast. Wetland-adapted.
Seeds contain 46% oil, 71% of which is petroselinic acid.

Source: Adapted from Draper (25), a comprehensive compilation of information on oilseeds of
potential economic value.

that precursors of specialty nylons might be grown in the
Southeast.

The plants of Table 4 would seem to be amenable to
low-cost, labor-intensive cultivation, and therefore, to be
of potential value for rural development in a part of the
United States that has long struggled with agricultural
depression. If new oilseed species are carefully chosen for
the Southeast to: (a) provide inherently valuable specialty
chemical products, (b) thrive on marginal lands without
heavy fertilizer or other inputs, and (c) to avoid the need
for expensive machinery for cultivation and harvesting, it
is possible that small-holder farming will be benefitted in
this region.

Table 4. Novel Oilseeds Potentially Cultivatable for
Specialty Chemical Crops in North Carolina (25)

Name	Comment
Stokesia laevis[b] *Euphorbia lagascae* *Crepis aurea* *Crepis vesicaria*[a]	Oil contains epoxy acids, for epoxy resins and related polymer products
Crepis alpina *Crepis foetida*[c] *Crepis rubra* *Lapsana communis*[a] *Picris hieracioides*[a] *Pyrularia pubera*[a,b]	Oil contains crepenynic acid, for low-temperature lubricants, polymers
Sapium sebiferum[c] *Sebastiana ligustrina*[c] *Leonotis nepetaefolia*[b,c] *Lamium purpureum*[a,b]	Oils are potential raw materials for bactericides, polymers, synthetic ester lubricants, new drying oils, new fatty chemical intermediates.
Eruca sativa *Iberis amara* *Iberis umbellata* *Thlaspi perfoliatum*[a,b]	Oils contain erucic acid for Nylon 13/13, plasticizers, elastomers, wax esters similar to sperm whale oil

(a) Found wild in North Carolina mountains.

(b) Found wild in North Carolina Piedmont.

(c) Found wild in North Carolina Coastal Plain.

Marginal-Land Oilseeds for Developing Nations

Potential new uses for oilseeds in developing nations include all those mentioned above for the United States (principally diesel fuel, chemicals for domestic use and export, and conversion to gasoline), plus fuel for cooking and lighting, mainly in rural areas. Of special concern is the shortage of firewood and charcoal in many developing nations, and the high cost of kerosene. Where oilseeds are inedible or can be grown on marginal lands unsuitable for food production, food versus fuel conflicts might be minimized. Marginal lands of all types are found in abundance in many developing nations, and support thousands of indigenous oilseed plant and tree species. Many of these are prolific producers of high-oil-content seeds that are presently underutilized or even ignored. In addition, many novel oilseeds might be adapted.

A consideration of adaptable marginal-land oilseeds for developing nations might include the buffalo gourd of the southwest United States. The cultivation of buffalo gourd in semi-arid regions of the world, beyond southwestern United States, is yet to be demonstrated. Early field studies in Lebanon gave promising results (18), but apparently there are no other published reports of trials in other developing nations that have hot-dry lands.

In an earlier study (19), we identified some East African oilseeds of possible economic value for fuels or chemicals, screened from a lengthy list provided by L. H. Princen of the Northern Regional Research Center, U.S. Department of Agriculture, Peoria, Illinois. Selections were made on the basis of published information on growth characteristics, habitat, and potential value of the oils. This preliminary effort to identify promising oilseeds included some that are found on marginal lands, e.g., Leucas martinicensus, a weed of disturbed soils of Kenya and other tropical countries, and Vernonia pauciflora, a Kenyan annual growing from dry bushlands to montane forests. Table 5 is a revised and expanded list of promising East African oilseeds. This is not comprehensive or exhaustive; many more entries could have been made.

In a similar manner (19), we identified Indian oilseeds of possible value, from another list supplied by L. H. Princen. Again, many are marginal-land plants, e.g., Azadirachta indica, Holarrhena antidysenterica, Maytenus emarginata, Portulaca quadrifida, and Wrightia tinctoria are indigenous dry-land plants, and Actinodaphne hookeri is a tree that is native to hilly, lateritic soils. Table 6 is a

Table 5. Potential New Oilseed Crops for East Africa

A. Drought-adapted plants

Azadirachta indica (Neem). Seeds 33-45% oil. Native to India, it thrives in dry nutrient poor areas. Has great potential as a source of firewood, for erosion control, and as an insecticide. Resistant to the desert locust. Neem has been planted in Zanzibar.

Balanites aegyptica (Lalob). Seed kernels contain 49% oil. A thorny savanna tree which is found in East Africa, Arabia, and the Sudan. Fruits resemble a date.

Moringa oleifera (Horse Radish Tree). Seed kernels contain 30-49% oil. A bush of poor soil in Kenya, North Africa, India and other areas in the tropics.

Salvadora persica. Seeds 34% oil. An evergreen shrub that grows on saline and sandy soils in savanna, along desert flood plains, and coastal alkaline savannas.

Vernonia pauciflora. Seeds 40-42% oil. An annual herb of Kenya growing over a wide ecological range from dry bushlands to montane forests.

Ximenia americana (Tallowwood, Wild Plum). Seeds 62.4% oil; oil contains an acetylenic acid (trans-11-octadecen-9-ynoic acid). A deciduous tree that is pantropical in distribution, growing in the Coast, Rift Valley, and Northern Provinces of Kenya as well as in India, Southeast Asia, Central and South America, and southern Florida along the coast. May be a root parasite.

Table 5, continued

B. Other plants

Calophyllum inophyllum (Indian Laurel). Seeds 60% oil. An evergreen tree of tropical coasts, including the East African coasts. Resists salt spray. Yields fruit in 3 to 4 years after seeds are planted.

Excoecaria bussei. Seeds 50-60% oil. A tree of north coastal Kenya.

Leucas martinicensus (Wild Tea Bush). Seeds 30-40% oil. A perennial herb which is weedy in disturbed soils of Kenya and elsewhere in the Tropics.

Telfairia pedata (Oyster Nut). Seeds 60% oil. A perennial vine with fleshy roots, native to lowland rain forests along the coast of Tanzania and Mozambique. Cultivated in Kenya. Vines grow to 100 feet in length, producing abundant fleshy, squash-like fruits.

Source: Adapted from Draper (25), a comprehensive compilation of information on oilseeds of potential economic value. Unpublished compositional information on E. bussei and L. martinicensus were provided by L. H. Princen from data obtained as part of the New Crops Screening Program, Horticultural and Special Crops Laboratory, Northern Regional Research Center, USDA, Peoria, Illinois.

Table 6. Potential New Oilseed Crops for the Indian Sub-Continent

A. Drought-adapted plants

Aegle marmelos (Bael Tree). 34.4% oil in seeds. A thorny deciduous tree of dry habitats, especially in central and south India, where it grows on dry slopes and plains. It is also cultivated in northwest India.

Azadirachta indica (Neem). 33-45% oil in seeds. Native to India, it thrives in dry nutrient-poor tropical and subtropical areas. It has great potential as a source of firewood, an oilseed for erosion control, and as an insecticide. It is resistant to the desert locust.

Balanites roxburghii (Hingot). Seeds contain 43% oil. Roots and fruits contain diosgenin. A small spiny tree that grows wild in the Indian desert.

Citrullus colocynthis (Thumba). Seeds contain up to 33% oil. A perennial cucurbit that grows wild and is cultivated in the Thar desert on sandy soils and sand dunes.

Crepis thompsonii. Seeds contain 16% oil, 65% of which is crepenynic acid. An annual of northwest India and Afghanistan.

Holarrhena antidysenterica (Kuda). Seeds 30-40% oil, 70% of which hydroxy fatty acids. A shrub or small tree of the semi-arid scrub forests of north India.

Maytenus emarginata (Kankero). Seeds 30-40% oil. A shrub of dry habitats in the Indian desert.

Portulaca quadrifida (Ram Jata). Seeds 60-70% oil. A succulent herb of the desert that spreads rapidly in fields and gardens.

Table 6, continued

Salvadora oleoides (Khakan). Seeds 42-43% oil, 47% of which is lauric acid. A tree or shrub that thrives in North India on semi-arid lands where rainfall is below 25 inches per year.

Wrightia tinctoria (Bhakar Oak). Seeds 30% oil. A shrub of semi-arid areas at desert edge, and also grows in other dry parts of India.

B. Tree crops for other marginal lands

Actinodaphne hookeri (Pisa). Seeds 63-82% oil, up to 96% of which is lauric acid. A tree of the eastern and western Ghats, growing on lateritic soil. A prolific fruit and seed producer.

Aphanamixis polystachya (Amoora). Seed kernels give 47% oil. An evergreen tree of India and southeast Asia, with potential for northeast India.

Bassia latifolia (Mahua). Seed kernels are 47-62% oil. An oilseed tree of southern India.

Calophyllum inophyllum (Indian Laurel, Punna, Undi). Seeds 60% oil. An evergreen tree of coastal southern India which resists salt spray. It yields fruit in 3 to 4 years after seeds are planted.

Mallotus philippinensis (Kamala). Seeds 32% oil. Seed oil is a good substitute for tung oil. Fruits give a commercially important dye in India. An evergreen tree that is abundant in sub-Himalayan areas of Uttar Pradesh.

Table 6, continued

Pongamia pinnata (Derris indica) (Poonga-oil Tree). Seeds contain 27-39% oil, which can be used as an illuminant or as a lubricant. This deciduous legume has been planted throughout the humid lowland tropics of the world, and it also adapts to drier parts of India. It is highly tolerant of salinity.

Putranjiva roxburghii. Seeds 40-50% oil. An evergreen tree. Seeds are used to produce a burning oil.

Santalum album (Sandalwood). Seeds 44-55% oil. A timber tree in India which produces two crops of seeds per year. This small evergreen tree is an obligate root parasite of other plants and must have suitable hosts in the vicinity to survive. It is economically important in Karnataka and Tamil Nadu for its essential oil from stems and roots.

Schleichera oleosa (Kusum, Gum Lac). Seeds 59-72% oil. A timber tree of deciduous forests. A tree that is scattered across the Ghats and along the east coast of India.

Sterculia foetida (Indian Almond, Java Olive). Seeds 44-54% oil. A quick-growing tropical shade and ornamental tree that will also tolerate salt spray. Its seeds contain 72% of a 19-carbon fatty acid with a cyclopropane ring.

Source: Adapted from Draper (25), a comprehensive compilation of information on oilseeds of potential economic value. Unpublished compositional information on A. hookeri, H. antidysenterica, M. emarginata, P. quadrifida and P. roxburghii were provided by L. H. Princen from data obtained as part of the New Crops Screening Program, Horticultural and Special Crops Laboratory, Northern Regional Research Center, USDA, Peoria, Illinois.

revised and expanded version of the earlier list, but is
still not exhaustive.

To the best of our knowledge, the use of high-oil-
content seeds as fuels is not yet a matter of practice.
Therefore, our research group has been interested in determining if the concept has value. Laboratory tests by
Mathieu (20) and cookstove tests by Schmidt (21) have now
shown that crushed oilseeds formed into cakes have promise
as novel fuels for Third World households. In addition,
lighting tests conducted by Mathieu (20) indicate that more
research and development is needed to determine if seed oils
might replace kerosene for lighting, using simple lamps of
appropriate design. Chapter 12 contains details of these
studies.

In addition to fuel uses, the chemical farming possibilities for developing-nation oilseeds should be given
close attention. For example, oil from <u>Actinodaphne hookeri</u>
is high in lauric acid, <u>Leonotis nepetaefolia</u> is a source of
allenic acids, and <u>Ximenia americana</u> and <u>Santalum album</u>
(Sandalwood) seeds yield acetylenic acids. Sandalwood, an
important timber tree in the tropics, produces two crops of
seed per year and the seeds are apparently not used for any
purpose, at present (19).

The Need for Evaluation of Marginal-Land Oilseeds

The economic benefits and the ecological risks of novel
oilseed cultivation should be evaluated, concurrently. The
need for new agronomic practices for various types of marginal lands should be analyzed, and where needed, such
practices should be developed. Special attention should be
given to the risks of large-scale oilseed farming on the
most fragile marginal-land ecosystems. On the positive
side, it is possible that oilseeds could be a step toward a
more permanent agriculture that could help stabilize the
soil. Annual oilseeds in which all vegetative residues were
returned might be more beneficial to soils than other proposed biomass energy schemes, such as widespread cultivation
of <u>Euphorbia lathyris,</u> which remove residues and, therefore,
risk losses of soil fertility. By comparison, perennial
oilseed crops such as trees could be beneficial to soil
fertility since there is little soil disturbance.

On the negative side, experience indicates that fertility of marginal lands, in general, often declines rapidly
after only a few years of cultivation unless correct cropping
practices are followed closely. This may enhance the competition for good cropland between food production and

oilseed fuel production, resulting in the serious ethical
and equity problems of "food versus fuel" that have already
surrounded the development of alcohol fuel from food crops.
Some oilseeds that will produce on marginal lands might
produce even better on prime farmland. Therefore, if the
market for an industrial crop became very attractive, it may
be impossible to prevent diversion of some good farmland
from food to fuel or chemical production. On the other
hand, many marginal-land oilseeds are not practical crops
for prime farmland. An example is the buffalo gourd which
requires dry-land growing conditions.

Even if the food versus fuel issue were minimized by
fuel and chemical farming only on marginal lands, another
issue of equally serious proportions would probably have to
be confronted, namely, the question of appropriate cropping
practices to preserve the stability of marginal lands. It
is not clear that agronomists will quickly turn their attention from prime farmland cropping problems to marginal-land
problems, nor is it obvious that government or other funding
will be easily available to support such agronomic and
ecological research and development work where needed. Even
if satisfactory marginal-land practices were available "on-
the-shelf," it is doubtful if marginal-land farmers would
adopt them quickly especially if the methods were unfamiliar
and required extra time and cost. If adoption of the best
available cropping practices for marginal lands occurs too
slowly, and the long-term stability and value of marginal
lands is thereby put at risk, this may or may not be recognized as an area for legitimate intervention by government
to control the way the land is cropped, in the national
interest. If intervention were acceptable, a number of
methods might be considered, including regulations and
penalties against inadequate practices, and incentives for
the adoption of adequate practices.

Another issue that needs to be analyzed is concerned
with insect pest populations in large-scale farming of new
oilseeds. Introduction of new crops would diversify the
agricultural base of a region, and this would generally help
to make the region less vulnerable to disruption by pests.
Already, some regions of the U.S. are associated with a
single crop, e.g., the Midwest with corn, and North Carolina
and Kentucky with tobacco. In 1970, the epidemic of southern
corn leaf blight caused losses that disrupted the agricultural economy of the South and Midwest (22). In 1979 and
1980, a blue-mold epidemic caused tremendous economic losses
of tobacco crops in the U.S., while in Cuba, blue mold set
in motion a chain of events that may have contributed to the
exodus of 125,000 Cubans to the United States (23).

While oilseeds might help to prevent agricultural disasters through crop diversification, it is important to be aware of potential pest problems associated with oilseeds as new crops, especially in intensive monoculture over large acreages. For example, Gmelina arborea is an Indian timber tree that was thought to be disease resistant. Large acreages were planted at Jari in Brazil. However, it grew poorly on sandy soils in the area and developed serious insect pest problems (24).

Other impact areas deserving of study include water-resource utilization, land-use patterns, food prices and supply, prevailing patterns of distribution of diesel and other petroleum-based fuels, and prices and supply of pesticides and fertilizer.

Conclusions

The high price of petroleum and the scarcity of firewood are factors causing a global search for alternative materials for fuels and chemicals. Oilseeds deserve consideration for these purposes, and offer many exciting opportunities for research and development, not only on the technical factors but also the economic, ecological, social and political aspects of these new crop introduction possibilities. Important global questions that must be addressed involve domestic issues such as the need to encourage domestic renewable production of strategic materials, and international issues such as the need for developed nations to assist developing nations to become more self-sufficient in renewable fuels.

Because prime farmland is mainly needed for food crops, oilseeds for fuels and chemicals should be encouraged on marginal lands. Many oilseeds suitable for such dry, wet, hilly, nutrient-poor, or saline soils are known, and might be domesticated and widely adopted, globally. Economic benefits and ecological risks when fragile marginal-land ecosystems are opened to agriculture should be analyzed and evaluated carefully. Satisfactory marginal-land cropping systems will very likely have to be developed, and marginal-land farmers encouraged or required to utilize such systems to prevent soil degradation and effective loss of the value of the land to future generations.

Many problems remain to be solved: technical, economic and sociocultural. The novel oilseed literature in key areas is incomplete in oilseed agronomy, oil extraction, conversion and utilization, especially for less well-known species. Further, the consequences of widespread novel

oilseed cropping are not clear in any part of the world.
Oilseeds may be promising supplements or replacements for
exhaustible, nonrenewable resources, but they are not
without their problems. Land-use issues, food-fuel
tradeoffs, economic questions, and the long-term stability
of marginal lands will continue to demand attention.

References and Notes

1. R.P. Morgan and E.B. Shultz, Jr., "Fuels and Chemicals
   from Novel Seed Oils, "Chemical and Engineering News 59,
   (September 7, 1981) 69-77.

2. From data reported by W.M. Potts, Chemurgic Digest 5,
   (1946) 373, 375, generally supported by Scheld's field
   investigations (Chapter 6).

3. P.G. Young, R.P. Morgan and E.B. Shultz, Jr., "Buffalo
   Gourd: Potential as a Fuel Resource on Semi-Arid
   Lands," Proceedings of the International Conference
   on Plant and Vegetable Oils As Fuels, ASAE Publication
   4-82, American Society of Agricultural Engineers,
   St. Joseph, MI (August, 1982).

4. Based on an estimated biennial yield of 13,500 kg of
   starch per hectare reported by Bemis (Reference 5b)
   and one gallon of ethanol per 15 pounds of starch,
   from Fuel From Farms: A Guide to Small-Scale Ethanol
   Production, Solar Energy Research Institute, Report
   No. SERI/SP-451-519 (February, 1980).

5. The following is a sampling of the literature on the
   buffalo gourd from the University of Arizona group
   headed by W.P. Bemis:

      a) W.P. Bemis, L.D. Curtis, C.W. Weber and
          J. Berry, "The Feral Buffalo Gourd, Cucurbita
          foetidissima," Economic Botany 32 (Jan-Mar
          1978) 87-95.

      b) W.P. Bemis, J.W. Berry and C.W. Weber, "The
          Buffalo Gourd, A New Potential Horticultural
          Crop," HortScience 13, (June 1978) 235-40.

      c) W.P. Bemis, J.W. Berry and C.W. Weber, "The
          Buffalo Gourd, A Potential Arid Land Crop,"
          in G.A. Ritchie, ed., New Agricultural Crops,
          (Westview Press, Boulder, CO, 1979).

      d) J.A. Vasconcellos, W.P. Bemis, J.W. Berry and

C.W. Weber, "The Buffalo Gourd, Cucurbita foetidissima HBK, as a Source of Edible Oil," in E.H. Pryde, L.H. Princen, K.D. Mukherjee, eds., New Sources of Fats and Oils, (American Oil Chemists' Society, Champaign, IL, 1981).

6. S.1462, U.S. Senate, July 10, 1981; H.R.4213, U.S. House of Representatives, July 21, 1981.

7. W.P. Bemis, J.W. Berry, M.J. Kennedy, D. Woods, M. Moran and A.J. Deutschman, Jr., "Oil Composition of Cucurbita," J. Amer. Oil Chemists' Soc. 44 (1967) 429-30.

8. W.P. Bemis, J.C. Scheerens, J.W. Berry, M.L. Dreher and C.W. Weber, "Accumulation of Crude Protein and Oil Contents," J. Amer. Oil Chemists' Soc. 54, (1977) 537-538.

9. A.S. Barclay, H.S. Gentry and Quentin Jones, "The Search for New Industrial Crops II: Lesquerella (Cruciferae) as a Source of New Oilseeds," Econ. Bot. (1962) 95-100.

10. H.S. Gentry and A.S. Barclay, "The Search for New Industrial Crops III: Prospectus of Lesquerella fendleri," Econ. Bot. 16, (1962) 206-211.

11. E.M. Apen, Jr., W.C. Cooper, R.J.M. Horton and L.D. Scheel, "Health Aspects of Castor Bean Dust," U.S. Dept. of HEW, Public Health Service Doc. AP-36 (1967).

12. R.J. Youle and A.H.C. Huang, "Evidence that the Castor Bean Allergens are the Albumin Storage Proteins in the Protein Bodies of Castor Bean," Plant Physiology 61 (1978) 1040-42.

13. H.J. Nieschlag, J.A. Rothfus, V.E. Sohns and R.B. Perkins, Jr., "Nylon-1313 from Brassylic Acid," Ind. Eng. Chem. Prod. Res. Dev. 16, (1977) 101-107.

14. We estimate that the total energy yield from buffalo gourd is over 40 million Btu per acre per year, including about 34 million Btu as ethanol, based on 400 gallons per acre per year, and at least 10 million Btu as seed oil, based on 2 barrels per acre per year. E. lathyris might yield 55 million Btu per acre per year if yield expectations (Reference 14a) could be achieved, but there is now serious doubt that this can happen (References 14b, 14c).

a)   J.D. Johnson and D.W. Hinman, "Oils and
     Rubbers from Arid Land Plants," Science 208,
     (May 2, 1980) 460.

b)   R.S. Loomis, "Agriculture," in L.E. St.-Pierre
     and G.R. Brown, eds., Future Sources of
     Organic Raw Materials, CHEMRAWN I, (Pergamon
     Press, 1980).

c)   R.F. Ward, "Euphorbia - Is it the Source of
     Hydrocarbons in the Future," Solar Energy 29,
     (1982) 83-86.

15. H.W. Scheld, N.B. Bell, G.N. Cameron, J.R. Cowles,
     C.R. Engler, A.D. Krikorian and E.B. Shultz, Jr., "The
     Chinese Tallow Tree as a Cash and Petroleum-Substitute
     Crop," In Tree Crops for Energy Co-Production on Farms,
     SERI Bull. CP-622-1086, (1981) pp. 97-111.

16. H.W. Scheld and J.R. Cowles, "Woody Biomass Potential
     of the Chinese Tallow Tree," Econ. Bot. 35, (1981)
     391-97.

17. C.L. Shih, Untitled paper in Chinese on the Chinese
     tallow tree. Research Bulletin of the Chekiang
     (Zhejiang) Forestry Research Institute, People's
     Republic of China, (1973).

18. L.C. Curtis, "An Attempt to Domesticate a Wild,
     Perennial, Xerophytic Gourd, Cucurbita foetidissima,"
     Progress Reports I-IV, (The Ford Foundation, 1972.)

19. E.B. Shultz, Jr., R.P. Morgan and H.M. Draper, III,
     "Oilseeds for Energy in Rural Areas of Developing
     Countries," Proceedings of the Symposium of the Inter-
     national Association for Advancement of Appropriate
     Technology for Developing Countries, Denver, CO
     (Oct., 1980).

20. Sandra L. Mathieu. "Potential Utilization of Oilseeds
     for Household Energy at the Village Level," M.S. Thesis,
     Department of Technology and Human Affairs, Washington
     University, St. Louis, MO (May, 1982).

21. E.W. Schmidt, S.L. Mathieu and E.B. Shultz, Jr.,
     "Comparison of Oilseed Fuels with Conventional Fuels
     in Simple Cookstoves," Paper presented at the 5th
     National Conference on the Third World, University
     of Nebraska at Omaha, Omaha, Nebraska, Oct. 27-30,
     1982.

22. G.W. Cox and Michael D. Atkins, <u>Agricultural Ecology</u>, (W.H. Freeman Co., San Francisco, 1979) pp. 520-522.

23. G.B. Lucas, "The War Against Blue Mold," <u>Science 210</u>, (1980) 147-153.

24. P.M. Fearnside and J.M. Ranken, "Jari and Development in the Brazilian Amazon," <u>Interciencia 5</u>, (1980) 146-156.

25. H.M. Draper, III. "New Oilseed Crops for Fuels and Chemicals: Ecological and Agricultural Considerations," DSc Dissertation, Dept. of Technology and Human Affairs, School of Engineering and Applied Science, Washington University, St. Louis, MO (December, 1982).

3. Endangered Plant Species: Preservation, Utilization, or Extinction?

I have been engaged in research on a familiar group of plants, the evening primroses (Oenothera) for more than 20 years, with the support of the National Science Foundation. The evening primroses are wild plants, mostly with yellow or white flowers, that occur in natural places and along roadsides throughout the United States. Some 60 kinds, or about half of the world total, occur in our country. Until recently, they have been prized mainly for their attractive flowers, which, for the most part, open in the evening.

Four of the evening primroses, the Eureka Dunes evening primrose (Oenothera avita subspecies eurekensis), the Idaho Dune evening primrose (Oenothera psammophila), the Arkansas suncups (Oenothera pilosella subspecies sessilis), and the Antioch Dunes evening primrose (Oenothera deltoides subspecies howellii) are considered endangered in the Smithsonian Institution's 1978 publication of "Endangered and Threatened Plants of the United States." Only the last-mentioned species has actually been listed as endangered by the Fish and Wildlife Service, and its critical habitat designated on the dunes at Antioch, California. In addition to these four evening primroses, an additional species, Oenothera organensis of the Organ Mountains of New Mexico, is listed by the Smithsonian Institution as threatened. Our own recent unpublished studies have indicated that a sixth species, Wolf's evening primrose (Oenothera wolfii) of the coastal dunes of northern California, also ought to be considered endangered.

Adapted from Testimony to The House Committee on Merchant Marine and Fisheries, Subcommittee on Fisheries and Wildlife Conservation and the Environment Oversight Hearings on Endangered Species Act, February 22, 1982.

The evening primroses might, on the basis of all knowledge available until about five years ago, simply have been
considered wildflowers and, as such, curiosities upon which
human welfare most certainly did not depend. Very quietly,
however, during the 1970's, commercial research began on
evening primroses in the Netherlands, Germany, and England --
all countries where weed evening primroses introduced from
the United States abound. The reason that these neglected
plants were gradually proving of interest to giant chemical
concerns in Europe was the discovery that the oil in their
seeds is one of the only two known rich natural sources of a
nutrient called gamma-linolenic acid (GLA). The other
natural source in which GLA is abundant is human milk. GLA
is a polyunsaturated substance which is also an essential
fatty acid. Essential fatty acids form part of the membranes
that surround the cells of the body; they are essential for
the proper functioning of these membranes. They are also
precursors of prostaglandins, which are hormones that are
produced by every organ of the body and control the second-
by-second regulation of organ functions.

It now appears that modern human populations are characterized by a widespread deficiency of essential fatty
acids -- a deficiency that seems to lead to many diseases
that are common in our population, including eczema, diseases
of the arteries, and arthritis. All of them are considered
mysterious in origin, and all are resistant to therapy. It
so happens that GLA is the most active of all essential fatty
acids in correcting these deficiencies. In other words, oil
derived from the seeds of these wildflowers may prove to play
an essential role in helping us to avoid coronary heart
disease and to cure such diseases as eczema and arthritis,
diseases that afflict millions of people in the U.S.
Research in medical schools all over the world is suggesting
many additional applications for GLA, and the scientific
press in the winter of 1982 has seen a proliferation of articles on the topic. Investors all over the world are suddenly
trying to find the best ways to convert a wayside wildflower
into a large-scale commercial crop.

Who knows which of the evening primroses of the United
States may prove to provide the richest source of gamma-
linolenic acid? The Antioch Dunes evening primrose was
federally listed primarily because it happened to occur in a
locality where there were two species of endangered butterflies. If the butterflies were not there, development of
the dunes at Antioch might have continued, and instead of
allowing a chemically unknown member of a group of plants
that produces a chemical that has now proved to be of intense

interest to human beings to continue to exist, we might
simply have had more cement manufactured from the beautiful
white dunes of Antioch. Would that have been progress, and
if so, for whom would it have been progress? It surely
brings to mind a memorable sentence from the World
Conservation Strategy recently developed by the International
Union for the Conservation of Nature and Natural Resources:
"We have not inherited the Earth from our parents, we have
borrowed it from our children."

The Endangered Species Act

The importance of a strong Endangered Species Act in the
United States cannot be overemphasized, not only because we
have a rich, beautiful, and valuable assemblage of plants and
animals within our borders, but because we historically have
been, and still are, leaders in the world conservation movement. Although we comprise less than one-thirtieth of all of
the people in the world -- and our proportionate representation is shrinking with every passing year -- we, as a nation,
have played a key role in giving rise to the conservation
movement, a movement in which our leadership is needed now
more than ever. Our Endangered Species Act, as well as the
Convention on International Trade in Endangered Species
(CITES), which were developed as integral parts of the same
plan, together help to provide the mechanism by which human
beings can preserve some of their options for the future.
Half a century after our great early conservationist Aldo
Leopold pointed out that the first rule of intelligent
tinkering was to save all the cogs and wheels, some of us
still have not come to appreciate the lesson. If we ignore
their individual importance to us, we may regard species
simply as dispensable impediments to whatever actions might
seem to be called for by the demands of the moment, instead
of as unique, and therefore priceless, elements in the
world's array of living things.

Plants, animals, and microorganisms working together in
complex interrelationships that are still very poorly understood make up the biosphere, and the worldwide web of life
of which we human beings are a part. As we progressively
modify this biosphere to cultivate our crops, grow our animals, and produce products of direct economic interest to us,
we simplify the relationships and increase the instability of
the system as a whole. As our actions promote the extinction
of organisms worldwide, we lose the "cogs and wheels" of
which Aldo Leopold spoke -- the elements which, like the
evening primroses, might have proved later to have been of
the greatest interest and importance to our descendants.

Worldwide Extinction of Plants

The extinction of plants and animals worldwide is probably, as my colleague Professor E. O. Wilson has pointed out,
the most significant event that is occurring in the world
during our lifetime. We know relatively little about many
species of tropical plants that are being lost as the world's
tropical rain forests are cut down. Only about one-sixth of
an estimated 3 million tropical plant and animal species
have been cataloged(1). In Latin America, about ten to fifteen thousand plants and trees have not been described scientifically. Furthermore, our knowledge of both ecological
processes and organisms in the tropics is meager. Newly discovered tropical plants might prove to be important sources
of food, fuel, medicine, and chemical feedstocks for the
rapidly increasing population of the Third World. More support is needed for basic tropical biology studies.

To the current world population of about 4.5 billion
people, roughly two billion more -- a number equal to the
entire population of the world in the year 1930 -- will be
added during the next 20 years. Some 90 percent of this
growth will take place in the tropics, where a majority of
the species of plants and animals, and by far the most poorly
known of these, occur. The World Bank has estimated that
some 800 million people in the tropics live in absolute pov-
erty at the present time, and there is little likelihood of
reducing this number, even proportionately, while the populations of tropical countries double -- inevitably because of
their age structure -- over the next quarter century. The
effects of this rapidly growing population on tropical vegetation cannot be overestimated, and many of us believe that
something like a million species, amounting to about a
quarter of the diversity of life on earth, will become
extinct during the next 30 years or so -- in other words,
within the lifetime of a majority of those alive at the
present day.

Many of the organisms that will become extinct in the
near future might have considerable economic potential, and
yet we are losing them so rapidly that we will not, in many
cases, have the opportunity to explore this potential. The
legumes, for example, are a large group of plants comprising
some 18,000 species. The members of this group are well
known for their ability to fix nitrogen, and thus make it
available for the enrichment of soils. Among the legumes are
many economic plants, such as peas, beans, soybeans, and
alfalfa, as well as many of the newly popularized fast-
growing tropical trees -- trees that hold great promise in
the desperate search for firewood that is increasingly

characteristic of the tropics. Which of the legumes can
safely be consigned to extinction? Or which of the grasses,
with about 10,000 species, including rice, wheat, rye,
barley, oats, corn, bamboos, and many other plants of tremendous economic importance, can we do without? Almost half of
the medical prescriptions written in the United States contain one product or more of natural origin. Furthermore,
many significant world crops have been only recently cultivated. For example, within a fifty-year period, the oil palm
has become the basis for a multibillion dollar industry.

Whether we think of the biosphere as a gigantic worldwide resource for human exploitation and modification, or
whether we think about the individual plants, animals, and
microorganisms that make up the biosphere as scientifically
or aesthetically interesting or potentially useful for human
welfare, the consequences of extinction can only be seen as
catastrophic. We must, as a human race, try to find ways to
ameliorate the consequences of this extinction, and the
United States and similar developed countries must continue
to provide the powerful role of leadership that they have
exercised in the past in this critically important area.

Preservation of Species Diversity

The importance of preserving species diversity is too
great to let the process of listing species for protection
continue to crawl slowly forward, encumbered by various non-
biological considerations. That process must be expedited
and based solely upon biological considerations so that we
can know when tradeoffs are being made and not be blind to
them. In addition, plants that have been federally determined to be endangered should be accorded the same protection
as animals, and it should not be legal to take them either
without specific permit procedures. There should be a resto-
ration of funds for selective studies of endangered and
threatened species in the individual states, and financial
support should also be accorded to foreign countries in this
area to the extent possible.

I would like to conclude, as a botanist, by emphasizing
the inversion of values that must come to characterize our
efforts to preserve endangered species, presumably because of
the sentimentality we feel towards such large vertebrates as
the California condor, the golden eagle, and the grizzly
bear. Plants, and only plants, are able to transform the
energy ultimately derived from the sun into a form in which
it can be used by living organisms. The several million
kinds of animals and microorganisms that exist in the world,
and make up the biosphere, owe their continued existence to

no more than 300,000 kinds of green land plants and algae
which have the ability to capture the sun's energy. Since 15
or more kinds of animals and microorganisms exist for every
single kind of plant, it may be assumed that the extinction
of one kind of plant may, in turn, ultimately bring about the
extinction of a dozen or even many more kinds of animals and
microorganisms. In addition, I hope that the example of the
evening primroses will have reminded us of the fact that
plants are themselves natural biochemical factories from
which we derive many important commercial products. We have
not even begun to investigate the great majority of plants
for any property of potential interest, and the chemicals
they contain are just beginning to be explored.

From either of these points of view, it makes obvious
good sense to hold on to the plants that we have. In the
United States, about 10 percent of the roughly 20,000 species
of plants should probably be classified as "endangered" or
"threatened" under the meaning of the Act, whereas roughly
half of the approximately 2,200 native plant species found in
Hawaii deserve such classification. The Smithsonian
Institution, in 1978, listed 1,485 kinds of plants as endangered, 1,408 as threatened, and 360 as extinct. During the
eight years in which the Endangered Species Act has been in
operation, the Department of Fish and Wildlife has listed 63
plant species as endangered or threatened, or about 2 percent
of the total. Many of the remainder will be lost if the
process is not accelerated.

The importance of the loss of genetic diversity on a
world scale was clearly brought out in the Strategy
Conference on Biological Diversity convened at the State
Department on November 16-18, 1981. As James L. Buckley,
then Under Secretary of State for Security Assistance,
Science, and Technology, brought out in his opening remarks,
we are permitting high rates of extinction to limit the
potential growth of biological knowledge, and thus limiting
options not only for ourselves, but for future generations.
Wild relatives of domesticated plants and animals of obvious
commercial importance should receive special attention: for
example, there are over 100 wild species of plants in the
United States and its territories that are in fact threatened or endangered and are the wild relatives of crops of
economic importance. It certainly does not require an elab-
orate argument to indicate why the preservation of these
plants is important. A wild perennial relative of corn,
recently discovered in Mexico and consisting of a few thousand plants on a hillside in the state of Jalisco, is a close
relative of, and interfertile with, corn, which is cultivated
over some 70 million acres in the United States -- an area

the size of Arizona -- where about a million farmers grow 7
billion bushels a year, valued at well over $20 billion.
Obviously, the loss of such a wild plant has clear and immediate significance, and can easily be identified as detrimental to human interests.

We have fossil evidence that 70,000 years ago our ances-
tors in the Middle East used the flowers of oriental poppies,
as well as those of other plants, to decorate their graves.
Most of us know the large scarlet flowers of these poppies
from our grandmother's gardens, where we saw and admired them
as children. Within the last decade, we have come to know
that these poppies contain a chemical known as thebaine, a
compound that can be transformed into the medically important
codeine easily, but can be converted into the highly abused
heroin only by prohibitively complex and expensive processes.
Since thebaine causes convulsions at low dosage, its abuse
potential is negligible. Oriental poppies, therefore, can be
grown commercially to replace the cultivation of opium
poppies that have contributed to such enormous problems
throughout the world and here in our own country. They are
now being grown commercially for this purpose in France, the
Netherlands, Japan, Israel, Yugoslavia, and probably Turkey
and the Soviet Union.

As a human race, therefore, we have known and admired
the beautiful flowers of oriental poppies for tens of thousands of years; only within the past few years have we come
to understand their agricultural and economic importance.
Who can speak to the potential economic importance of the
millions of species of plants and animals that coexist with
us now, and who is wise enough to decree that any one of them
should be consigned to extinction? We do owe it to those who
will come after us to "save the cogs and wheels," because we
have learned only a small part of what we need to know about
intelligent tinkering yet. Hopefully, we will have the wis-
dom to preserve as many options for survival as possible for
ourselves, our children, and our grandchildren.

References

1. P. H. Raven, "Tropical Rain Forests: A Global
Responsibility," Natural History, 90, No. 2 (February 1981)
28-31.

Everett H. Pryde

4. Chemicals and Fuels from Commercial Oilseed Crops

Introduction

For hundreds of years, oilseed crops and animal fats have supplied societal needs for both food calories and technological applications, e.g., axle greases, lubricants, light sources, protective coatings, and soap. They will continue to do so for many hundreds of years into the future, but we must make certain that their utilization is to the maximum benefit to society, with appropriate balance between food and non-food uses. In the recent past history of the United States, this balance has been maintained at a food:nonfood ratio of 2:1 (1).

Not only must an appropriate balance be maintained, but also optimum technological utilization must be assured because fats and oils are limited, although renewable, resources. Limitations are those imposed by land availability and crop productivity. In the following discussion, production and markets for fats and oils in general as well as for vegetable oils in particular will be reviewed as an introduction to possible future uses for vegetable oils as alternative chemical feedstocks or as farm fuels.

Fats and Oils

Of the 9 million-metric-ton fats and oils industry in the United States, one-half is consumed in edible oil products, one-quarter goes into exports, and the remaining quarter is used in nonfood products (2). Nonfood utilization of fats and oils corresponds to only 2 percent of the total synthetic organic chemical production from petrochemicals, but the time is fast approaching when increasing amounts of alternative chemicals may need to be prepared from fats and

Table 1. World and U.S. Production of Fats and Oils[a]

Fat or Oil	World Production		U.S. Production	
	1971[b]	1980/81[c]	1980/81[c]	Percent of world production
	1,000 Metric Tons			
Edible vegetable oil				
Corn	289	518	373	72
Cottonseed	2,654	3,241	655	20
Olive	1,437	1,840	--	--
Peanut	3,377	3,003	99	3
Rapeseed	2,508	3,838	--	--
Safflowerseed	226	255	36	14
Sesameseed	721	611	--	--
Soybean	6,266	12,278	8,115	66
Sunflowerseed	3,612	4,695	607	13
Babassu	72	130	--	--
Coconut	2,514	3,325	--	--
Palm	1,937	5,034	--	--
Palm kernel	462	696	--	--
Total	26,075	39,464	9,885	25
Industrial				
Castor	346	389	--	--
Linseed	1,236	696	69	10
Oiticica	20	14	--	--
Olive residue	131	172	--	--

Table 1. Continued

Fat or Oil	World Production		U.S. Production	
	1971[b]	1980/81[c]	1980/81[c]	Percent of world production
Tung	141	90	---	--
Total	1,874	1,361	69	5
Marine oils				
Fish	1,173	1,187	100	8
Whale	70	10	---	--
Whale, sperm	135	58	---	--
Total	1,378	1,255	100	8
Animal Fats				
Butter (fat content)	4,114	4,927	445	9
Lard	4,421	3,827	454	12
Tallow and grease	4,568	6,047	3,351	55
Total	13,103	14,801	4,250	
Fats and oils, overall total	42,430	56,881	14,304	29

(a) Calculated from assumed crushing rates applied to that portion of each crop available for crushing and/or export and, therefore, represents potential not actual production.
(b) Reference 4.
(c) Reference 5, 6.

Table 2. World Production and Exports of Fats and Oils, [a,b]
1978.

	Amount, 1,000 Metric Tons	
	U.S.	World
Production	30,555	52,575
Exports	14,824	18,436
Exports, percent of production	49	35

[a] Reference 2.

[b] Includes fat or oil equivalent of oilseeds exported for crushing and of butter.

oils. Animal fats now contribute the major share (58 percent) to fat-derived industrial products, with vegetable oils, both edible and industrial, supplying the balance. Vegetable oils, however, will be increasing their total nonfood markets as the economy grows and as animal fat production levels off (1-3).

The United States produces about 25 percent of the total world production of edible vegetable oils, 5 percent of total industrial vegetable oils, 8 percent of marine oils and 29 percent of animal fats (Table 1). The United States exports more of its vegetable oil production than most countries, about one-half of its total production compared to an average of about 35 percent for the world (Table 2). The United States also imports about 900 thousand metric tons of fats and oils, mainly coconut and palm oils (Table 3). About one-half of the coconut oil imported is used for producing detergent and other nonfood products.

During the "age of petroleum," fats and oils retained, perhaps to a surprising degree, many of their traditional markets; this can be attributed in part to their useful properties and in part to the fact that many are byproducts of other industries and, as a result, have been priced competitively with petrochemicals. Because petrochemical prices have increased so rapidly since 1973 while fats and oils have increased their price at more moderate rates, the latter materials have improved their competitive position. Table 4 indicates the share of the total market for various nonfood products which fats and oils provided in 1974 (1). If these markets continue to grow at the rates expected, and if fats and oils maintain the share indicated, an additional 1.35 million metric tons of fats and oils will be needed by 1990, doubling the amount used in 1978. If fats and oils increase their share of markets (2), which is a likely circumstance, as much as 3.4 million metric tons over and above the present volume of 1.36 million metric tons could be required by 1990.

Vegetable Oils

Commercial seed oils, including herbaceous and tree crops, contribute more than 450,000 metric tons annually to the industrial materials economy in the United States (Table 5). Surprisingly, edible oils contribute almost twice as many tons as inedible oils. The total vegetable oil contribution to industrial products corresponds to about 0.8 percent of the petrochemical industry or about 5 percent of the total fats and oils industry and about 20 percent of

Table 3. U.S. Production, Imports, Domestic Disappearance,[a] and Exports of Selected Fats and Oils (7)[b,c]

Oil or Fat	Volume, 1,000 Metric Tons			
	Production	Imports	Domestic Disappearance	Exports
Food Vegetable Oil				
Coconut	---	506	488	14
Corn	303	5	264	42
Cottonseed	518	---	243	277
Palm	---	300	277	26
Palm kernel	---	71	50	---
Peanut	165	---	105	21
Safflower	---	---	16	---
Soybean	3,890	---	3,381	729
Sunflower	6	---	---	---
Animal Fat				
Lard	479	---	367	113
Tallow, edible	241	---	242	10
Tallow, inedible and grease	2,469	3	1,442	1,292
Industrial Vegetable oil				
Castor	---	54	49	---
Linseed	98	---	74	6

Table 3. Continued

| Oil or Fat | Volume, 1,000 Metric Tons | | | |
	Production	Imports	Domestic Disappearance	Exports
Tall oil	536	---	460	80
Tung	---	10	9	---
Vegetable oil foots[d]	---	---	34	---
Marine Animal Oil				
Fish, sperm, and mammal oil	61	4	25	53
Total[e]	8,766	953	7,526	2,663

(a) Domestic disappearance is equivalent to sales, losses, and other items of consumption.

(b) Exclusive of butter fat and oil equivalent of exported oilseeds.

(c) For crop year 1976/77. Crop years vary from commodity to commodity and are defined in Reference (7).

(d) Vegetable oil foots includes the soapstock recovered from edible oil refining processes.

(e) The totals given represent minimum values since all-inclusive data are not available.

Table 4. Percent of Market Share for Nonfood Products Provided by Fats and Oils in 1978 (2)

Product	Percent
Adhesives	1
Agrichemicals	10
Coatings	40
Engineering thermoplastics	2
Fabric softeners	90
Plastics additives	15
Surfactants	45
Synthetic lubricants	20

Table 5. Consumption of Seed Oils in Inedible Products in the United States (8)

Seed oil	Year			
	1977	1978	1978/79[a]	1979/80[a]
	1,000 Metric Tons			
Edible:				
Coconut	244.7	239.1	219.0	185.6
Corn	0.4	0.5	---	1.0
Cottonseed	4.6	2.8	2.5	3.8
Palm	5.9	5.2	10.3	12.8
Peanut	1.2	1.4	1.0	0.7
Safflower	2.4	2.0	2.0	1.4
Soybean	105.9	112.5	110.5	93.0
Subtotal	345.0	363.5	345.3	298.3
Inedible:				
Castor	49.4	47.5	50.3	40.8
Linseed	97.2	103.5	94.1	72.6
Vegetable oil foots	39.1	43.1	39.1	36.4
Subtotal	185.7	194.1	183.5	149.8
Other	32.3	32.8	32.7	---
Overall total	563.0	590.4	561.5	448.1

[a] Crop year beginning October 1 and ending September 30.

the fats and oils going to industrial products. The edible oils going into industrial products consist mainly of coconut and soybean oils. The inedible oils consist mainly of castor and linseed oils as well as soapstocks obtained from the refining of edible oils.

The United States produces two-thirds of the world's soybean oil (Table 1). As a consequence of poor economic conditions, hardening of the dollar relative to other currencies and softening of export markets, a record 770,000 metric tons (1.7 billion pounds--about 2 months' supply) of soybean oil has been carried over from the 1980/81 crop year to the 1981/82 crop year (9). The result has been depressed prices for vegetable oils generally as well as for soybean oil. Since November 1981, crude soybean oil has been available below 20 cents per pound.

Other seed oils of major importance to the fats and oils industry produced in the United States include cottonseed, sunflower, corn and peanut oils (Table 1). Sunflower oil has received a great deal of attention in recent years, resulting in increased production, particularly in the Dakotas and Minnesota, and in the construction of several oil extraction plants. Soybean oil contributes the greatest amount to industrial products, with markets such as alkyd resins (172 million pounds in 1977) and epoxidized plasticizer/stabilizers for vinyl plastics (87 million pounds in 1980).

Alternative Chemicals from Vegetable Oils

As stated previously, we must be certain that the most beneficial uses of vegetable oils in both food and nonfood applications are considered in future planning. Vegetable oils are valuable renewable resources but may have limited future availability. For the present, vegetable oils are readily available and their prices are depressed, but as petrochemical prices continue to increase, greater demands will be made upon vegetable oils for feedstock. For example, several new plants have been built or are under construction to produce detergent fatty alcohols from vegetable oils rather than from petrochemicals. There can be little doubt that detergents fill a justifiable need, but there may be other technological needs that have higher priorities. Whatever the circumstance, new technologies need to be developed continuously. Products that could form the basis for new technologies and that are based upon research carried out at the Northern Regional Research Center (NRRC) include:

(a) Vinyl ethers. Fatty vinyl ethers of linseed alcohols can be either polymerized alone to adherent, chemically resistant coatings or copolymerized with styrene to make coatings of various degrees of flexibility as required by the specific end use (10).

(b) Cyclic fatty acids. Alkali treatment of linseed oil causes cyclization of linolenic acid. The cyclic acids are potentially useful as lubricants (11,12) or in alkyd resins (13). Other types of cyclic acids can be made from linseed fatty acids (14) or soybean soapstock (15).

(c) Fatty aldehydes have been made by reductive ozonolytic cleavage of unsaturated fatty compounds (16,17) or by the addition of carbon monoxide and hydrogen to unsaturated fatty acids by the hydroformylation (oxo) reaction (18). The aldehyde group is a versatile one that may be easily converted to acetals, alcohols, carboxylic acids and amines, which have potential value in polymers, plastics and plastics additives.

(d) Acetals of aldehyde esters have potential value as plasticizers, whether from 9-carbon (19) or 19-carbon (20) aldehyde esters.

(e) Amino acids have potential value for nylon-9 engineering thermoplastics (21) as well as for polyamide resin adhesives and printing inks (22).

(f) Polyols formed by reaction of formaldehyde with 9-carbon aldehydes (23) or 19-carbon aldehydic acids (24) may have use in coatings, lubricants and plasticizers with improved flexibility and water resistance.

(g) Polycarboxylic acids have possibilities in coatings, lubricants, plasticizers and polyamide resins (18,25).

(h) Organosulfur compound research will lead to a better understanding of the sulfurization reaction and to better extreme-pressure lubricant additives (26) to take the place of products derived from sperm whale oil.

(i) Organosilicon research is still undergoing basic investigations, although some results of NRRC-sponsored research are already available (27). It is probably too early to speculate on future potential uses.

Special attention should be directed toward potential coatings and plasticizer uses. Alkyd resins for coatings are the largest nonfood market for soybean oil, but this is an aging technology in an era of restrictions on atmospheric pollution by paint solvents. Two approaches have been investigated at NRRC to dispense with solvents; one approach involves high-solids baked coatings for metals (28), and the other involves energy-conserving, water-dispersible, alkyd-type resins that dry rapidly and have excellent flexibility (29).

The other area of major importance is in plasticizers for vinyl plastics. Dioctyl phthalate, the most commercially important plasticizer, has become a ubiquitous environmental contaminant and has been accused of being a possible health hazard (30). This situation may provide an excellent opportunity for vegetable oil derivatives to fill a pressing need. There are several vegetable oil-based plasticizers that could serve in place of dioctyl phthalate (31-33).

Fuel Uses of Vegetable Oils

A few years ago, a temporary shortage of diesel fuel caused some problems in crop production. The farmer has only limited time "windows" during which he can plant or harvest and cannot afford to wait even a few days for fuel during such times. Accordingly, much interest has been generated in vegetable oils as an emergency alternative fuel, and a considerable number of engine tests have been initiated to evaluate the oils (see Chapter 9). Short-term engine performance tests have given optimistic results; longer term endurance tests have given variable results.

Vegetable oils for diesel fuel have a number of advantages. They are liquid fuels from renewable resources and have a favorable energy input/output ratio, unless produced on irrigated land. They would permit crop production even in a petroleum shut-off and have potential for making marginal lands productive. They consume less energy than does alcohol production and have higher energy content than alcohol. They have cleaner emissions and simpler technology than alcohol production. One disadvantage is that vegetable oils as yet are not economically feasible ($1.50 to $2.00 per gallon vs. 1.25 per gallon for diesel oil). Further research and development are needed. On-farm processing technology has not yet been developed completely.

Vegetable oils can be used successfully in a naturally aspirated, air-cooled, indirect-injection diesel engine (34); they cannot be used neat in direct-injection engines. However, sunflower oil in a 20-percent solution in diesel oil apparently can be used in a direct-injection engine. Simple esters of the vegetable oils, which have much lower viscosity and better volatility than the oils, apparently perform well in the direct-injection engine, but long-term endurance tests have not yet been completed. The problem is that the great majority of farm tractors in the United States have direct-injection engines for greater fuel efficiency; therefore, some kind of modification to vegetable oils appears to be necessary before they can be used in this type of equipment.

Can vegetable oils supply all farm energy needs? Probably not. United States on-farm diesel oil requirements in 1978 were 3.3 billion gallons, up from 2.8 billion gallons in 1974. The amount of vegetable oils required to substitute for the 3.3 billion gallons is on the order of 28 billion pounds, larger than the entire fats and oils industry in the U.S. This statement is based on the present pattern of production for oilseed crops. Scientists at North Dakota State University and in the Republic of South Africa point out that, if the farmer devoted about 10 percent of his land to sunflower, the oil produced would be sufficient to operate the rest of the farm. This amount represents the penalty (tithing, if you will) that society must pay to maintain high technology in crop production. To abandon high technology would place society in a far more serious--and probably catastrophic--situation.

To develop basic information on both physical and chemical modification of vegetable oils needed to improve their properties as fuels, the Northern Regional Research Center has undertaken investigations on aqueous ethanol/vegetable oil microemulsions and on the transesterification reaction of simple alcohols with vegetable oils.

The vegetable oil microemulsions are based on studies previously carried out on microemulsions of aqueous alcohol as a fuel extender for both gasoline and diesel oil (35,36). Both ionic and nonionic microemulsions are being studied (37).

The incorporation of aqueous ethanol into vegetable oil as a microemulsion serves not only to extend diesel fuel supplies but possibly also to improve combustion properties. Vegetable oils, when injected into the combustion chamber of a diesel engine cylinder, do not form the "atomized" spray typical of No. 2 diesel oil. As a consequence, combustion is incomplete, and injector coking, ring sticking, and

lubricant contamination are major problems. We believe that
the microemulsions will form a better spray pattern and give
superior engine performance because of the lower viscosities
of the microemulsions compared to the original oil. Initial
engine tests being conducted at the University of Illinois
by Professor C. E. Goering are encouraging. Evidence is
being accumulated in our laboratory that the nonionic
formulation is indeed a microemulsion near the plait point
of the ternary diagrams that we have developed.

In our study of the transesterification reaction, we
have developed a quantitative method of analysis for the
reaction products using the Iatroscan TH-10, which employs
thin-layer chromatography and a flame ionization detector
(38). With a molar ratio of 4.2 to 1 of methanol to soybean
oil (40 percent excess methanol), the conversion of soybean
oil to its methyl esters exceeded 90 percent. With lesser
amounts of methanol, considerable quantities of mono- and
di-glycerides were produced. The presence of these components
may be undesirable, since their viscosities are higher than
the original oil and they are more apt to produce acrolein
in the exhaust gases of a diesel engine.

Potential for Oilseed Crop Production

The 360 million acres (146 million hectares) of cropland
in use represents 77% of the 470 million acres (190 million
hectares) of total cropland available in the United States.
Corn (34×10^6 hectares), soybeans (28×10^6 hectares) and
wheat (33×10^6 hectares) occupied more than half of the
available cropland in 1980. Hay production at 24×10^6
hectares also is a major item. The acreage devoted to
oilseed crops in the world as well as in the United States
is shown in Table 6. Most of the corn and the greater part
of the soybeans (as defatted meal) go into animal feeds in
the United States. Soybean oil is really a minor coproduct
of soybean meal, just as cottonseed oil is a byproduct of
cotton production.

Oil contents and average yields for different oilseed
crops are listed in Table 7. Because of its high productivity,
corn produces almost as much oil per acre as soybeans, in
spite of its low oil content. There are corn varieties that
have twice as much oil and still retain high productivity.
However, corn oil is a byproduct of the starch industry,
which consumes only a small portion of the corn produced,
and was produced only to the extent of 373,000 metric tons
for the 1980/81 crop year (Table 1). Nevertheless, a much
higher potential for corn oil production exists.

Table 6. U.S. vs. World Plantings in Oilseed Crops

Oilseed crop	World Planting Area 1972/73-1976/77[a] average	World Planting Area 1980/81[b]	U.S. Planting Area 1980/81[b]	U.S. Planting Percent of world planting
	1,000 hectares			
Cottonseed	31,970	32,583	5,348	16
Flaxseed	5,530	5,111	285	6
Peanuts	18,430	17,599	566	3
Soybeans	39,088	50,156	27,461	55
Sunflowerseed	9,124	12,103	1,568	13

[a] Reference 39.
[b] Reference 4.

Table 7. Oil Content and Average Oil Yield for Some Oil Crops

Oil crop (location)	Oil content, wt. %[a]	Average oil yield, kg/ha[b,c]
Palm oil (Malaysia)[d]	20	3,475
Copra (Phillipines)	65-68	800
Peanuts (U.S.)	45-50	790
Safflower (U.S.)	30-35	762
Sunflower (U.S.)	40-45	589
Rapeseed (Canada)	40-45	409
Soybean (U.S.)	18-19	319
Corn kernel (U.S.)	4.8	254
Flaxseed (U.S.)	35-42	230
Sesame (India)	45-50	220
Cottonseed (U.S.)	18-20	140

[a] Reference 40.
[b] Based on 1970-74 averages, except sunflower, which is based on 1977 and 1978 data because of switch to hybrids.
[c] Reference 41.
[d] An additional 420 kg/ha was obtained from the palm kernel, which contains 45-50 percent oil (palm kernel oil).

Soybean oil, although present in only 18-19 percent amounts in the bean, dominates the vegetable oil industry to such an extent that there is a record amount of carryover. However, if a real need should arise for more vegetable oil as feedstock or fuel, then more emphasis needs to be placed on high-oil crops such as safflower, sunflower, and peanuts. Peanuts can be bred to even higher oil content, and the high-oil varieties can be genetically marked so that they can be distinguished easily from edible varieties. The coconut and oil palm trees can be grown only in limited areas of the United States, but they produce large quantities of oil, particularly the oil palm. Coconut oil is a valuable source for both fuel and feedstock. Other potential oilseed crops that have high oil contents have been discussed elsewhere in this volume and in a recently published monograph (42).

Overview

Vegetable oils probably will be available for fuel and alternative chemical use only to a limited extent. On the one hand, vegetable oils can not be expected to be the sole alternative to petroleum products and petrochemicals. On the other hand, vegetable oils can be expected to make partial and significant contributions to these areas in the future. Vegetable oil supplies are overstocked at present. Pending outcome of engine evaluation tests over the next several years, one method of disposing of these stocks would be to burn them in diesel oil blends, or in an emergency, to burn them neat. Another method of disposing of supplies would be to develop new but elastic alternative chemical markets on the order of 45 to 227 thousand metric tons (100 to 500 million pounds). Such markets would help stabilize prices for vegetable oils to benefit both farmer and consumer. As has been shown for corn markets, even as little as 2 percent of total corn demand, operating at the margin in the market, is a price-strengthening force (43).

The overall impression is that vegetable oils can and will supply a small proportion of society's needs in other than edible oil products. Should the demand keep increasing, there are a number of alternative oilseed crops now under development that will satisfy the demand.

References

1. E. H. Pryde, "Nonfood Uses for Commercial Vegetable Oil Crops," in Crop Resources, D. S. Seigler, ed., Academic Press, Inc., New York, N.Y. (1977) 25-45.
2. E. H. Pryde, "Fats and Oils as Chemical Intermediates: Present and Future Uses," J. Am. Oil Chem. Soc., 56 (1979) 849-854.
3. E. H. Pryde, "Vegetable Oil Raw Materials," J. Am. Oil Chem. Soc., 56 (1979) 719A-725A.
4. U.S. Department of Agriculture, Foreign Agricultural Service, Foreign Agriculture Circular (Oilseeds and Products) FOP 17-81, The Service, Washington, D.C. (October 1981) 6-8.
5. U.S. Department of Agriculture, Foreign Agricultural Service, Foreign Agriculture Report FOP 25-77, The Service, Washington, D.C. (December 1977) 9.
6. P. Mackie, "World Oilseeds Outlook," in Foreign Agricultural Circular--Oilseeds and Products, FOP 20-81. U.S. Department of Agriculture, Foreign Agricultural Service, Washington, D.C. (November 1981) 23-37.
7. U.S. Department of Agriculture, Economics, Statistics and Cooperatives Service, Fats and Oils Situation, FOS-293, the Department, Washington, D.C. (October 1978) 23-24.
8. U.S. Department of Commerce, Bureau of the Census, Current Industrial Reports, Fats and Oils, Production, Consumption and Factory and Warehouse Stocks, Report Nos. M20K(78)-13, M20K(79)-13, and M20K(80)-13, Washington, D.C., 1979, 1980, and 1981.
9. U.S. Department of Agriculture, Economic Research Service, Fats and Oils, Outlook and Situation, FOS-305, Washington, D.C. (October 1981) 6.
10. B. G. Brand, H. O. Schoen, L. E. Gast, and J. C. Cowan, "Evaluation of Fatty Vinyl Ether Polymers and Styrenated Polymers for Metal Coatings," J. Am. Oil Chem. Soc., 41 (1964) 597-599.
11. J. P. Friedrich, E. W. Bell, and L. E. Gast, "Potential Synthetic Lubricants: Esters of C18-saturated Cyclic Acids," J. Am. Oil Chem. Soc., 42 (1965) 643-645.
12. J. P. Friedrich and R. E. Beal, "Liquid C18 Saturated Monocarboxylic Acids--Their Preparation, Characterization and Potential Uses," J. Am. Chem. Soc., 39(12) (1962) 528-533.
13. W. R. Miller, H. M. Teeter, A. W. Schwab, and J. C. Cowan, "Alkyd Resins Modified with Cyclic Fatty Acids. A Preliminary Evaluation," J. Am. Oil Chem. Soc., 39 (1962) 173-176.

14. E. W. Bell and L. E. Gast, "Alkyd Resins Modified with Tetrafluoroethylene Adduct of Conjugated Linseed Fatty Acids," J. Coat. Technol. 50(636) (1978) 81-87.

15. R. E. Beal, L. L. Lauderback, and J. R. Ford, "Soybean Soapstock Utilization: Fatty Acid Adducts with Ethylene and 1-Butene," J. Am. Oil Chem. Soc., 52 (1975) 400-403.

16. P. E. Throckmorton and E. H. Pryde, "Pilot Run, Plant Design, and Cost Analysis for Reductive Ozonolysis of Methyl Soyate," J. Am. Oil Chem. Soc., 49 (1972) 643-648.

17. E. H. Pryde and J. C. Cowan, "Ozonolysis," in Topics in Lipid Chemistry, F. D. Gunstone, ed., Logos Press Limited (now available from John Wiley and Sons, Inc., New York), Vol. 2, Chapter 1 (1971).

18. E. N. Frankel and E. H. Pryde, "Catalytic Hydroformylation and Hydrocarboxylation of Unsaturated Fatty Compounds," J. Am. Oil Chem. Soc. 54 (1977) 873A-881A.

19. E. H. Pryde, D. J. Moore, J. C. Cowan, W. E. Palm, and L. P. Witnauer, "Azelaaldehydic Acid Ester-acetal Derivatives as Plasticizers for Poly (Vinyl Chloride)," Polym. Eng. Sci. (1966) 60-65.

20. R. A. Awl, E. N. Frankel, E. H. Pryde, and J. C. Cowan, "Acetal Derivatives of Methyl 9(10)-Formylstearate: Plasticizers for PVC," J. Am. Oil Chem. Soc., 49 (1972) 222-228.

21. R. B. Perkins, Jr., J. J. Roden III, and E. H. Pryde, "Nylon-9 from Unsaturated Fatty Derivatives: Preparation and Characterization," J. Am. Oil Chem. Soc., 52 (1975) 473-477.

22. W. R. Miller, W. E. Neff, E. N. Frankel, and E. H. Pryde, "9(10)-Carboxyoctadecylamine and 9(10)-Aminomethyloctadecanoic Acid: Synthesis and Polymerization to Polyamides with Lateral Substitution," J. Am. Oil Chem. Soc., 51 (1974) 427-432.

23. D. J. Moore and E. H. Pryde, "Improved Synthesis of 1,1,1-Trimethylolalkanes from Hexanol and Nonanal. J. Am. Oil Chem. Soc., 45 (1968) 517-519.

24. W. R. Miller and E. H. Pryde, "9,9(10,10)-Bis(acetoxymethyl)octadecanoate Esters as Plasticizers for Poly(vinyl Chloride)," J. Am. Oil Chem. Soc., 55 (1978) 469-470.

25. W. L. Kohlhase, E. N. Frankel, and E. H. Pryde, "Polyamides from Carboxystearic Acid," J. Am. Oil Chem. Soc., 54 (1977) 506-510.

26. A. W. Schwab, L. E. Gast, and H. E. Kenney, "Tetrasulfide Extreme Pressure Lubricant Additives." U.S. Patent 4,218,332 (August 19, 1980).

27. N. Saghian and D. Gertner, "Polymerization of 1,3,5-tri(1,3,5,7-tetra)-methyl-1,3,5-tri(1,3,5,7-tetra)-10-carbomethoxydecylcyclotri(tetra) Siloxane," J. Macromol. Sci.-Chem. A9 (1975) 597-605.
28. F. L. Thomas and L. E. Gast, "New Solventless Polymeric Coatings from Fatty Acid Derivatives," J. Coat. Technol., 51(657) (1979) 51-59.
29. W. J. Schneider and L. E. Gast, "Poly(ester-amide-urethane) Water-dispersable and Emulsifiable Resins," J. Coat. Technol., 51(654) (1979) 53-57.
30. Anonymous, "A Phthalate Plasticizer Causes Cancer in Animals," Chem. Week (October 22, 1980) 14.
31. E. N. Frankel, W. E. Neff, F. L. Thomas, T. H. Khoe, E. H. Pryde, and G. R. Riser, "Acyl Esters from Oxo-derived Hydroxymethylstearates as Plasticizers for Poly(vinyl Chloride)," J. Am. Oil Chem. Soc., 52 (1975) 498-504.
32. E. J. Dufek, F. L. Thomas, E. N. Frankel, and G. R. Riser, "Some Esters of mono-, di-, and tricarboxystearic Acid as Plasticizers: Preparation and Evaluation," J. Am. Oil Chem. Soc., 53 (1976) 198-203.
33. W. R. Miller, E. H. Pryde, and G. R. Riser, "9,9(10,10)-Bis(acetoxymethyl) Octadecanoate Esters as Plasticizers for Poly(vinyl Chloride)," J. Am. Oil Chem. Soc., 55 (1978) 469-470.
34. F. Hugo, "Sunflower Oil as a Diesel Fuel Replacement: The South African Research Program," Proceedings, Vegetable Oils as Diesel Fuel--Seminar II, October 21-22, 1981, Northern Regional Research Center, Peoria, Illinois (November 1981).
35. P. A. Boruff, C. E. Goering, A. W. Schwab, and E. H. Pryde, "Engine Evaluation of Diesel Fuel-Aqueous Ethanol Microemulsions," Trans. ASAE, in press (Paper No. 80-1523 presented at the 1980 Winter Meeting, American Society of Agricultural Engineers).
36. A. W. Schwab, R. S. Fattore, and E. H. Pryde, "Diesel Fuel-Aqueous Ethanol Microemulsions," J. Dispersion Sci. Technol., 3 (1982) 45-60.
37. A. W. Schwab and E. H. Pryde, "Vegetable Oil Microemulsions as Diesel Fuel," Proceedings, Vegetable Oils as Diesel Fuel--Seminar II, October 21-22, 1981, Northern Regional Research Center, Peoria, Illinois (November 1981).
38. B. Freedman and E. H. Pryde, "Fatty Esters from Soybean Oil for Use as a Diesel Fuel," Proceedings, Vegetable Oils as Diesel Fuel--Seminar II, October 21-22, 1981, Northern Regional Research Center, Peoria, Illinois (November 1981).

39. U.S. Department of Agriculture, Foreign Agriculture Service, Foreign Agriculture Circular (Oilseeds and Products) FOP 24-79, The Service, Washington, D.C. (December 1979) 11.

40. E. H. Pryde and H. O. Doty, Jr., "World Fats and Oils Situation," in New Sources of Fats and Oils, E. H. Pryde, L. H. Princen, and K. D. Mukherjee, American Oil Chemists' Society, Champaign, Illinois, 1981, 1-14.

41. H. O. Doty, Jr., "U.S. and World Soybean Oil Markets," in Handbook of Soy Oil Processing and Utilization, D. R. Erickson, E. H. Pryde, O. L. Brekke, T. L. Mounts, and R. A. Falb, eds., American Oil Chemists' Society, Champaign, Illinois (1980) 485.

42. E. H. Pryde, L. H. Princen, and K. D. Mukherjee, eds., in New Sources of Fats and Oils. American Oil Chemists' Society, Champaign, Illinois, 1981.

43. C. A. Moore. High Fructose Corn Syrup: The Impact of a New Product on the Corn Economy. Unpublished report, Northern Regional Research Center, ARS/USDA, Peoria, Illinois, 1980.

Robert Kleiman, L. H. Princen,
Harold M. Draper III

5. Chemicals and Fuels from Wild Plant Oilseeds

Introduction

If the energy requirements for conversion are included, approximately 10% of our annual petroleum (or crude oil) demand is used for the production of plastics, coatings, lubricants, detergents, and a host of other petrochemical materials. The annual U.S. consumption of such petrochemicals is on the order of 100 billion pounds. Although it is generally believed that petrochemicals have taken over totally from earlier feedstocks, even today plant and animal products still play a prominent role in the manufacture of organic chemicals.

Traditional commercial fats and oils are very limited in their chemical constitution. All are triglycerides, made up predominantly from stearic, oleic, linoleic, and linolenic acids. In an effort to develop new oilseed crops for industry and farmers alike, the U.S. Department of Agriculture (USDA) started a screening program for which the chemical analyses were performed at the Northern Regional Research Center (NRRC) at Peoria, Illinois. During the past 23 years, between 7000 and 8000 species of wild plants have been analyzed for oil and protein content, fatty acid composition, and other chemical and physical characteristics (1-3). This program has resulted in the identification of about 75 new fatty acids never before described from any source, as well as many potential new crops that could be used as excellent sources for industrial lipids or fatty acids. Several have been studied genetically and agronomically, and a few have reached the stage of crop commercialization.

The mention of firm names or trade products does not imply that they are endorsed or recommended by the U.S. Department of Agriculture over other firms or similar products not mentioned.

71

The screening survey of the Northern Regional Research Center becomes even more important if one also considers the recent interest in developing alternative liquid fuels to replace diesel oil and other petroleum fractions for energy production. We have discovered many plant seeds that contain between 50 and 70% oil, which is much higher than any of the commercial crops now considered for such energy purposes. Many of these potential energy crops may not be suitable for the United States, but even so they may be excellent sources for heating and cooking energy in developing nations where firewood and other energy sources are in short supply.

Sources of Lauric Acid

World production of lauric acid-containing oils (coconut, palm kernel, and babassu kernel) totals some 7 million metric tons. Currently, no domestic source of this raw material exists. Price instability and political uncertainties make domestic production of such a crop inviting at this time. In 1960, Cuphea llavea was found to contain 83% capric acid (4). This discovery led to examination of other species in the genus and identification of species high in caprylic acid (C. hookeriana) and lauric acid (C. carthagenensis) (5). Recently, Graham et al. (6) reported the composition of 46 species of Cuphea, 21 of which are rich in lauric acid. The highest concentration found was 79% of the total fatty acids present. Although most Cuphea are herbaceous in habit, many agronomic problems must be solved before Cuphea can be a commercial crop for the United States. These agronomic problems include irregular germination, indeterminate flowering, stickiness to the touch, and seed shattering. Work is now underway in West Germany at the University of Göttingen to solve some of the shattering and stickiness problems encountered in wild Cuphea. We have found that Cuphea is very attractive chemically, with seed oil percentages as high as 34%. If agronomic problems can be eliminated, these species can be a valuable source of raw material for soaps, detergents, and lubricants.

Three Cuphea species native to the United States may be of interest as lauric acid crops. The most widespread in geographic range is Cuphea carthagenensis, which has 33% oil in its seeds (5) and 62.5% lauric acid in the seed oil (6). It is an annual species introduced from South America that is found in the southeastern coastal plain from North Carolina to eastern Texas and southern Florida. Its habitats are low wet meadows, sunny ditches, marshes, and disturbed places (7,8). The wet adaptation is of interest, because poor drainage and waterlogging are problems in many southern coastal plain and coastal flatwood soils.

Cuphea glutinosa is found in the western coastal plain region of Texas and Louisiana, in open woods and pastures (8). Its seed oil contains 53.6% lauric acid and 21.3% capric acid (6), and it is a short-lived perennial.

Cuphea wrightii is more semiarid in adaptation. An annual species with 54% lauric acid and 29.4% capric acid, it is found in canyons and on slopes of low hills between 4000 and 6000 feet in southeastern Arizona and southwestern New Mexico (9,10). Widespread in Mexico, its habitat is in the semiarid mountain ranges of Sonora and Chihuahua (11) extending south in the mountains to Puebla (12) and Mexico City (13). Cuphea wrightii is a summer-growing plant that blooms in August, surviving in areas that have 12 weeks of summer rains amounting to 15 inches, with little or no rain the rest of the year (11).

Some tropical and subtropical shrubs and trees of the family Lauraceae produce seeds that typically have large amounts of lauric acid in their oil. Only a relatively few of the 2500 species have been chemically explored, but these show potential as sources of lauric acid. Three examples include Actinodaphne hookeri, Neolitsea umbrosa, and Litsea cubeba.

Actinodaphne hookeri seeds were found to have 71% oil and 90% lauric acid. This dioecious tree is found in greatest abundance in the western Ghats of India. Seeds are produced in great abundance (14), and the Indian government lists it as one of the twelve most important nonedible oilseeds in India (15). It is abundant enough to support a wild harvest of 340 tons of seeds per year (16).

Neolitsea umbrosa is an evergreen shrub of the Himalayas found in high-elevation temperate climates (17). Seeds contain 66% oil and 59% lauric acid.

Litsea cubeba (66% oil and 59% lauric acid) is a Chinese tree of warm temperate regions, including the eastern Himalayas, Yunnan, and Shanghai (18,19). It can probably be cultivated in other warm temperate regions as well.

A native woody perennial of the United States, Lindera benzoin (spicebush), is a potential lauric acid source. Native to the eastern United States, the seed contains 61% oil and 47% lauric acid (42% is capric acid, also a valuable raw material). L. benzoin is a deciduous shrub found in alluvial woods and along streams in both the northeastern and southeastern United States (7,20). Spicebush seeds are sown in the fall, and the expected germination percentage the following spring is 70-80% (21).

Although it is not the current trend to attempt to cultivate woody perennials as seed oil producers, these plants might have some distinct advantages over annual crops if mechanical harvesting methods could be developed.

Seed oils of the family Umbelliferae generally contain abundant amounts of petroselinic (cis-6-octadecenoic) acid (22). This acid can be cleaved at the double bond either through ozonolysis or other methods into lauric and adipic acids. Although we are concerned mainly about a source of lauric acid, the byproduct, adipic acid, which is now a petrochemical, could then be supplied partially from agriculture. The family includes many plants of economic importance. However, none is grown for its seed oil. We have found a number of plants in this family that produce seed with oil contents greater than 25%, and one as high as 50% (22). Foeniculum vulgare, common fennel, generally grown as a source of condiment and essential oils, contains 24% oil in its seed. This crop could be used as is or bred to a higher seed oil content for use of the petroselinic acid for subsequent lauric and adipic acid production. Foeniculum vulgare is an annual herb that grows up to 3 feet; the vegetation is used in salads, and the seed in confections. Seeds are sown in early spring and are available from many seed companies. One northeastern company, Comstock, Ferre and Co., recommends in their 1982 catalog that it be grown in the southeastern United States "especially, but not exclusively."

Oils Containing Hydroxy Acids

Castor oil is the only seed oil of commerce that contains a fatty acid with a hydroxy functional group, namely ricinoleic acid. A number of other seed oils have been found to have large amounts of hydroxy fatty acids. Three of these species, identified as having crop potential, differ somewhat from castor in the structure of the major hydroxy fatty acid.

Lesquerella spp. native to the U.S. generally contain several hydroxy fatty acids in major (50-70%) concentrations (23). One of these acids is the C_{20} homologue of ricinoleic acid, lesquerolic (14-hydroxy-cis-11-eicosenoic) acid. The same reactions applied to ricinoleic acid can be applied to lesquerolic acid. For example, the reaction of ricinoleic acid with sodium hydroxide under pressure produces capryl alcohol and sebacic acid, whereas under the same conditions, lesquerolic acid results in capryl alcohol and dodecanedioic acid. Dodecanedioic acid now is produced from nonrenewable resources.

There are 12 Lesquerella spp. with large percentages of lesquerolic acid in their seeds, and all but one of them are found in the southwestern United States. The southwestern species tend to be adapted to calcareous soils or to rather specific ecological conditions such as blackland prairies (24). Plants tend to be adapted to dry conditions, growing contiguously in colonies containing hundreds of thousands of individuals. This is especially true of Lesquerella fendleri, L. gracilis, and L. gordonii (24).

Lesquerella fendleri has been harvested by combine in a wild stand to give a yield of 1000 lb of seed per acre (25). Seeds contain 20-28% oil and 62% lesquerolic acid. The plant is found in the southern high plains, southeastern Arizona, southwestern Texas, and the Edwards Plateau of south central Texas.

Lesquerella gordonii occupies many of the same regions as L. fendleri, extending farther east in the Texas prairies and growing on sandy soils instead of calcic soils. Lesquerella gracilis is a prairie plant found on the Texas blackland prairies and grand prairies, as well as on the Cherokee prairies of Kansas and Oklahoma (24).

A southeastern species is Lesquerella globosa, which is adapted to limestone soils in Kentucky and Tennessee (24); its seeds contain 39% oil and 66% lesquerolic acid. This perennial was placed on the list of threatened plant species in the United States (26).

The seed oils from Dimorphotheca and some Osteospermum spp. are rich (up to 70%) in 9-hydroxy-trans,trans-10,12-octadecadienoic (dimorphecolic) acid (27). This multifunctional fatty acid can be converted to a conjugated trienoic acid similar to that found in tung oil, or can react with malic anhydride or other reactive dienophiles to produce chemical intermediates (28).

Two annual Dimorphotheca species of interest are D. sinuata and D. pluvialis. Dimorphotheca sinuata seed contains 42% oil, 65% of which is dimorphecolic acid, while D. pluvialis seeds contain 25-35% oil, with 60-75% dimorphecolic acid. These two species are native to the Mediterranean climate region of South Africa (29). In August and September, D. pluvialis covers the slopes of the Cape region with white flowers (30). Both species are weedy and would appear well-adapted to cultivation, except that they exhibit poor seed retention (29).

A perennial species with good seed retention is D. cuneata. It is found in droughty habitats in the Karroo desert of South Africa, and also in semiarid mountain areas that receive snow in the winter. Seeds are 32% oil, 69% of which is dimorphecolic acid (29). Estimated seed yields from a California planting were 1603-2615 lb/acre (31).

An Osteospermum species of interest is O. ecklonis, an evergreen shrub of Mediterranean climates in South Africa with a seed oil percentage of 50%, 67% of which is dimorphecolic acid (29). It is planted extensively in southern California as a ground cover along highways (32). Potential seed yields are 1350 lb/acre (31).

Both 9,10,18-trihydroxyoctadecanoic and 9,10,18-trihydroxy-cis-12-octadecenoic acids can be isolated from the seed oil of Ptilostemon afer (Chamaepeuce afra). These polyfunctional molecules could be the basis of diverse chemical reactions. Ptilostemon afer is a biennial of thistle-like habit, found in rocky stony areas of the Balkan peninsula (33) and Turkey (34). It thrives in soil rich in lime (33).

Seed Oils Containing Epoxy Fatty Acids

Seed oils of species from a number of botanical families have been identified to have large amounts (up to 80%) of monoepoxy fatty acids in their seed oils (35). These include the Compositae, Euphorbiaceae, and Linaceae. Two species of the Compositae, Vernonia galamensis and Stokesia laevis (both with about 40% oil and 75% epoxy fatty acids), are most promising agronomically. These epoxy oils can be used in coatings or as plasticizers. With an oxirane content of greater than 4%, they are similar to some commercially epoxidized oils and should be excellent ingredients for adhesives.

Native to East Africa, Vernonia galamensis is an annual herb found over a wide ecological range from dry bushland to montane forest (36). It has been grown in Puerto Rico and Kenya (37). It requires short days, and therefore, may not do well in the United States (38).

Stokesia laevis, a perennial herb native to the southeastern United States, is widely cultivated in flower gardens throughout the eastern United States. It does best in well-drained soil under cultivation, even though its

native habitat is wetlands (39). Seeds are available from
seed companies in the U.S. A number of good agronomic
characteristics make it attractive as a crop. In addition
to yields up to 2000 kg/ha, the seeds are above the foliage,
which facilitates harvest. Stands are productive for 3 to
5 years, reducing the need for annual planting. There is
good genetic variation. A disadvantage is that Stokesia
plants will not flower until the second year if seeds are
planted in the spring. Achene dormancy and self-
incompatibility inhibit breeding and agronomic work (40).

Oils with Unusual Unsaturation

At the Northern Regional Research Center, we have
identified many fatty acids with unusual olefinic and
acetylenic functional groups. As an illustration, Crepis
alpina seed oil contains about 70% cis-9-octadecen-12-ynoic
(crepenynic) acid (41). This unique structure presents
opportunity for cyclisation and, therefore, has potential as
a source of low-temperature lubricants.

Crepis alpina is a plant of Turkey and Syria (34) that
has been evaluated agronomically by the U.S. Department of
Agriculture (42). It can be grown as a winter annual or an
annual in the southern parts of the U.S., and can be harvested
by combine; it produces 1800 kg/ha of seed (42).

Alternative Fuel Source

Several commercial vegetable oils now are being
considered seriously as diesel oil replacements or for other
fuel uses. Our survey has identified seeds with exceptionally
high oil contents. For example, 50 plant species were found
to contain between 60 and 70% oil, and another 88 species
contain between 50 and 60% oil (43). These findings may
have special significance for third-world countries
(developing nations), where such seeds may be collected from
the wild and either pressed to produce liquid fuel or burned
directly as solid fuel.(See chapter 12).

Conclusion

We have presented an overview of the results generated
from the "New Crops" program of the Agricultural Research
Service. The program has generated several papers describing
the many species of plants with useful chemical constituents.
There still lies ahead the need for plant breeders and
agronomists to tame these wild plants so that in the future
we may call these species crops. In addition, American
industry might explore opportunities for converting their

facilities to use renewable resources and agriculturally
derived raw materials rather than depending upon oil and
coal for their supplies.

References

1. Q. Jones and F. R. Earle, Econ. Bot. 20 (1966) 127.
2. A. S. Barclay and F. R. Earle, Econ. Bot. 28 (1974)
    179.
3. L. H. Princen, J. Am. Oil Chem. Soc. 56 (1979) 845.
4. T. L. Wilson, T. K. Miwa, and C. R. Smith, Jr., J. Am.
    Oil Chem. Soc. 37 (1960) 675.
5. R. W. Miller, F. R. Earle, I. A. Wolff, and Q. Jones,
    J. Am. Oil Chem. Soc., 41 (1964) 279.
6. S. A. Graham, F. Hirsinger, and G. Röbbelen, Am. J.
    Bot. 68 (1981) 908.
7. A. E. Radford, H. E. Ahles, and C. R. Bell, "Manual of
    the Vascular Flora of the Carolinas," University of
    North Carolina Press, Chapel Hill (1968).
8. S. A. Graham, Sida 6 (1975) 80.
9. T. H. Kearney, and R. A. Peebles, "Flowering Plants and
    Ferns of Arizona," USDA Misc. Publ. (1942) 423.
10. W. C. Martin, and C. R. Hutchins. " A Flora of New
    Mexico," J. Cramer, Vaduz (1981).
11. H. S. Gentry, "Rio Mayo Plants," Carnegie Inst. of
    Wash. Publ. 527 (1942).
12. H. Puig, "Vegetation de la Huasteca, Mexique," Mission
    Archeologique et Ethnologique Francaise au Mexique,
    Mexico (1976).
13. S. O. Sanchez, "Flora del Valle de Mexico," Editorial
    Herrero, Mexico City (1969).
14. H. Santaupau, "The Flora of Khandala on the Western
    Ghats of India," Botanical Survey of India, Calcutta
    (1967).
15. Department of Science and Technology, Government of
    India. "Utilization and Recycling of Agricultural and
    Animal Wastes/By-Products; A Country Report," New Delhi
    (1977).
16. S. S. Wagle, A. M. Lele, and D. D. Kelkar, Indian
    Oilseeds J. 8 (1964) 1981.
17. "Wealth of India, Raw Materials," Vol. IX. Council of
    Scientific and Industrial Research, New Delhi (1972).
18. S. C. Lee, "Forest Botany of China," The Commercial
    Press, Shanghai (1935).
19. H. Hara, "The Flora of the Eastern Himalaya," University
    of Tokyo Press, Tokyo (1971).
20. S. A. Spongberg, J. Arnold Arboretum 56 (1975) 1.
21. K. A. Brinkman and H. M. Phipps. In "Seeds of Woody
    Plants in the United States," pp. 503-504, USDA Agric.
    Handb. 450 (1974).

22. R. Kleiman and G. F. Spencer, J. Am. Oil Chem. Soc. 59 (1982) 29.
23. R. Kleiman, G. F. Spencer, F. R. Earle, H. J. Nieschlag, and A. S. Barclay, Lipids 7 (1972) 660.
24. R. C. Rollins and E. A. Shaw, "The Genus Lesquerella (Cruciferae) in North America." Harvard U. Press, Cambridge, MA (1973).
25. H. S. Gentry and A. S. Barclay, Econ. Bot. 16 (1962) 206.
26. E. S. Ayensu and R. A. deFilipps, "Endangered and Threatened Plants of the United States." Smithsonian Institution and World Wildlife Fund, Washington (1978).
27. F. R. Earle, K. L. Mikolajczak, I. A. Wolff, and A. S. Barclay, J. Am. Oil Chem. Soc. 41 (1964) 345.
28. C. R. Smith, Jr., In "Fatty Acids," E. H. Pryde, ed., American Oil Chemists' Society, Champaign, IL (1979) 29-47.
29. A. S. Barclay and F. R. Earle, Econ. Bot. 19 (1965) 33.
30. M. R. Levyns, "A Guide to the Flora of the Cape Peninsula," Jutta and Company, Cape Town (1966).
31. B. C. Willingham and G. A. White, Econ. Bot. 27 (1973) 323.
32. H. S. Gill, G. A. Zentmyer, O. K. Ribeiro, and L. J. Klure, Plant Dis. Rep. 60 (1976) 647.
33. T. G. Tutin, V. H. Heywood, N. A. Burges, D. H. Moore, D. H. Valentine, S. M. Watters, and D. A. Webb, eds., "Flora Europaea," Cambridge Univ. Press, London (1972).
34. P. H. Davis, "Flora of Turkey," Edinburgh U. Press, Edinburgh (1975).
35. F. R. Earle, J. Am. Oil Chem. Soc. 47 (1970) 510.
36. A. D. Q. Agnew, "Upland Kenya Wild Flowers," Oxford U. Press, London (1974).
37. K. D. Carlson, W. J. Schneider, S. P. Chang, and L. H. Princen, In "New Sources of Fats and Oils," E. H. Pryde, L. H. Princen, and K. D. Mukherjee, eds., Am. Oil Chem. Soc., Champaign (1981) 297-317.
38. G. A. White, B. C. Willingham, W. H. Skrdla, J. H. Massey, J. J. Higgins, W. Calhoun, A. M. Davis, D. D. Dolan, and F. R. Earle, Econ. Bot. 25 (1971) 22.
39. G. A. White and C. R. Gunn, Garden J. 24 (1974) 84.
40. T. A. Campbell, In "New Sources of Fats and Oils," E. H. Pryde, L. H. Princen, and K. D. Mukherjee, eds., Am. Oil Chem. Soc., Champaign (1981) 287-295.
41. K. L. Mikolajczak, C. R. Smith, Jr., M. O. Bagby, and I. A. Wolff. J. Org. Chem. 29 (1964) 318.
42. G. A. White, B. C. Willingham, and W. Calhoun, Econ Bot. 27 (1973) 320.
43. R. Kleiman, L. H. Princen, and H. M. Draper III. In preparation.

H. W. Scheld, Joe R. Cowles, Cady R. Engler,
Robert Kleiman, Eugene B. Shultz, Jr.

6. Seeds of the Chinese Tallow Tree as a Source of Chemicals and Fuels

Introduction

The hazards of developing a new crop plant are many.
Markets for the products are uncertain. The chemical nature
of the new products and the technologies required for their
production and utilization are often imperfectly understood.
Furthermore, there is a near certainty of biological sur-
prises which may have considerable impact upon the agronomic
system to be utilized, if not upon the ultimate practicality
of developing the crop. Nevertheless, the currently changing world energy picture with a concomitant rise in value
of products that can serve as convenient forms of stored
energy or substitute for petroleum-derived chemicals seems
to provide a suitable climate for the launching of efforts
to develop new sources of industrial raw materials. Moreover, technological advances and accumulated knowledge
point emphatically toward the desirability of exploiting
more fully the potential of the world's botanical heritage.

Historically, perennial crop plants have played a
minor role in the U.S. agricultural economy, their main contribution being in non-staple food crops such as fruits and
nuts. Minor exceptions have been the naval stores industry
based upon the long-leafed southern yellow pine, and tung
oil from the tung tree, <u>Aleurites</u>. Because of labor costs,
the supply of terpenes from tapping of the pine tree has
declined drastically to be replaced by tall oil, a by-
product of the Kraft pulping of southern pine wood. The
once-thriving Gulf coast tung oil industry has long since
died and the few surviving orchards are being uprooted to be
replaced by annual crops. However, with the onset of high
energy costs, the need for exploitation of less-choice land
areas, and the growing concern over soil loss under modern
intensive cropping practices, it is timely to consider the
increased use of perennial crop systems. Such systems would

tend to have greatly reduced energy inputs because of their "one-time" establishment costs. Trees, especially, are able to exploit a large portion of their environment in terms of soil and air space; hence productivity is potentially high. Elimination of extensive annual tillage results in greatly reduced runoff and soil loss, and moreover, should allow the utilization of sensitive land areas which would be quickly destroyed by conventional annual cropping practices.

In recent years, workers in the U.S. have examined several uncultivated trees or shrubs for their crop potential. The most notable example has been the jojoba, Simmondsea chinensis, an oilseed-bearing shrub that appears to be on the way to success as an arid land specialty crop in a number of countries. Among other species that have been given serious consideration are the honey locust, Gleditsia triacanthos, and mesquite, Prosopis spp., both producers of carbohydrate-rich pods, and the Chinese tallow tree, Sapium sebiferum (L.) Roxb. (1, 2), a tree with a high yield of oilseeds. In this chapter we will attempt to convey a relatively complete picture of the Chinese tallow tree, its origins, characteristics, and potential, primarily as a perennial tree crop for the southern United States, and to give an indication of its possible usefulness in other parts of the world where labor-intensive cropping systems are practical. Although our primary concern is with the Chinese tallow tree as an oilseed crop, we have taken the view that the potential for multiple uses may have a very large influence upon economic viability for this principal use. Therefore, the possibilities for employing the Chinese tallow tree as a fuelwood crop (2), as a honey crop, and as a source of natural product extractives are also considered.

The Chinese tallow tree is poised to become a useful crop in the United States in part because it is already well along the path to development. The stock available in the southeastern U.S., though presently uncultivated, is nevertheless a relatively recent escape from stock which had already been subjected to several centuries of selective cultivation in China. While unsophisticated, the seed harvesting and processing technology is well established. Further, knowledge of the chemical characteristics of the seed fat and oil is now available.

The Chinese tallow tree is an introduced subtropical euphorb which has become naturalized in the lowlands of the Gulf and the South Atlantic coasts and in southern California. It is notable for its hardiness and high growth rate, for its ability to grow in poorly drained, saline soils, for its relative freedom from insect pests and pathogens in the United States, and for its ability to produce large quantities of oil and fat-rich seeds (Figure 1).

The seed is comprised of an outer coating of hard, saturated fat, embedded in a fibrous matrix, separated from an oily seed kernel by a hard impermeable shell. The kernel oil is highly unsaturated, and is a mixture of about two-thirds conventional triglycerides, and one-third tetraester triglycerides (estolides). The estolides contain two unusual short-chain fatty acids, 8-hydroxy-5,6-octadienoic acid (HODA) and trans-2,cis-4-decadienoic acid (DDA). The protein-rich seed meal has an attractive amino acid profile and is rich in thiamin (3).

Brief History of Culture and Use

The Chinese tallow tree is not a wild tree. It is a plant that has been selected, improved and cultivated by traditional Chinese methods for possibly a thousand years, and whose seed products have been known to commerce for at least one hundred years (4, 5). Traditionally, the tree appears to have been cultivated for much the same reasons as are now under consideration: as a productive crop on lands marginal for or unsuited to conventional crops. In the 19th century, MacGowan (4) observed that in China:

> "They are seldom planted where anything else
> can be conveniently cultivated - in detached
> places, in corners about houses, roads, canals,
> and fields."

Only in relatively recent times has there been an appreciable effort devoted to development of formal orchards (6, 7).

The uses to which the tree has been put are many and varied. The tallow (Chinese vegetable tallow), was used in candle and soap making, in the dressing of cloth and the sizing of paper, and as an edible fat (6, 8). It was exported from China for use in soap and candle making and to a limited extent as an edible fat (5, 9). The kernel oil (Stillingia oil) was used as an illuminating or lubricating oil, in paints and finishes, and cosmetically because it was believed to restore color to graying hair (4, 8, 10).

The press cake from oil extraction was used as cooking fuel or returned to the soil as a fertilizer, especially for tobacco (4). More recent reports suggest that both tallow

Figure 1.   Seeds of the Chinese tallow tree.

and oil are presently used extensively in the Chinese chemical industry (6, 7, 8, 10).

Various parts of the tree have been used in Chinese herbal medicine; it is considered to be one of the best of the "cool" (effective in treatment of fever) medicinal plants (11). Wu-Chiu-Keng, a drug described as purgative and diuretic, has been prepared from the root bark (4). A black dye for silk, probably a tannin, has been extracted from the leaves (12). The wood has been used in the manufacture of implement handles and small carved articles such as clogs and printing blocks.

Although the tree has repeatedly been considered as a crop plant (10, 12, 13-19), there is no record that it has ever been seriously or successfully cultivated anywhere except in its native China, perhaps because methods and equipment for culture and processing were somewhat more complex than for most oilseeds. Its introduction into the United States appears to have occurred in 1772, at the suggestion of Benjamin Franklin. When he was in London at that time, Franklin sent Chinese tallow tree seeds to Noble Jones of Georgia, who was apparently the first to plant the tree in the southeastern United States (20). Franklin had hoped that the Chinese tallow tree would prove to be an economic source of fats for candles and soap. The major systematic introduction was an effort in the early 20th century by the U.S. Bureau of Plant Industry to establish local soap industries in the south, the major localities being Jacksonville, Florida, and Houston, Texas (13). These attempts failed; nevertheless they added impetus to the spread of the tree which is now well established throughout the Gulf and lower Atlantic coasts.

Growth Habit, Morphology and Physiological Characteristics

The Chinese tallow tree is deciduous, seldom growing beyond a height of 40 feet, superficially resembling an aspen with oppositely or spirally arranged glossy, heart-shaped leaves which often produce brilliant fall color including yellow, orange, red, purple and bronze. The form ranges in different strains from low, spreading, and multi-forked to slender and columnar with small pendant branches. The bark is rough and gray; boles of older specimens become gnarled and fluted.

Form in the Chinese tallow tree is set by a tendency toward weak apical dominance. In open-grown trees the typical form exhibits multiple forks and a weakly dominant

terminal with large heavy branches. Dwarf trees in which
height never exceeds 15 feet are not uncommon. Branches
and twigs, especially in older trees, tend to be pendant.
Because the Chinese tallow tree is shade-intolerant, boles
grown in a closed stand under crowded conditions tend to be
unbranched but crooked as a result of continual change in
dominance of various branches with subsequent death of the
losers. Boles that do not lean or have a pronounced sweep
are rare. As a result of the growth habit, boles of older
trees are characteristically gnarled and eccentric in cross
section. Rarely, an individual tree will be found which
exhibits a tall form with a straight columnar trunk and
very small branches.

The pattern of growth in the Chinese tallow tree as we
observe it now in the United States is well suited to
orchard culture or to a short-rotation silvicultural system.
Initial height and diameter growth are extremely rapid, but
there is a physiological "shifting of gears" between 5 and
10 years. Annual increments of growth become so small and
diffuse that distinguishing them separately for estimation
of age is extremely difficult. Concurrently, terminal
growth declines greatly and in the United States the terminals tend to die back in the wintertime with a season's
flush of growth just replacing the loss due to death, with
crown size remaining static. In general, this shifting of
gears appears to be associated with a major change in the
way the tree partitions its photosynthate from vegetative
growth to seed production. The mechanisms governing the
shift are presently unknown; detailed study could shed useful light upon productivity of the Chinese tallow tree.

There are no printed reports, at least in the Western
literature, concerning the physiology of the tree and such
work has only recently begun at the University of Houston.
The descriptions here will thus be limited to a level of
gross observations. The physiological characteristics of
the Chinese tallow tree are those of a pioneer species and
one that is generally adapted to wet conditions. Growth is
rapid and sexual maturity is reached within a relatively
short time. The tree is shade-intolerant and shaded leaves
and branches are quickly lost.

Large catkins of yellow flowers are formed in spectacu-
lar array in the spring (May to June in the United States);
seeds are borne in clusters of green, usually 3-lobed
capsules which blacken, dry and split in the fall (September
to November in the United States) to expose white, pea-sized
seeds. Sexual maturity has been observed in the second
season of growth at least in the sense that small abortive

male flowers are formed in some of the terminals. In the populations studied thus far, approximately 50% of the population form flowers in the third season. Complete flowers are formed and seeds set by a small percentage of the populations in the third season of growth. At this time most of the trees that flower can be induced to produce complete flowers by application of stressing treatments such as reversal of a small band of bark. Flowering appears to be dependent not so much upon photoperiod as upon temperature. Trees separated by about 400 miles along the Texas coast exhibit a difference in flowering time of approximately 20 days. Flowering can be induced at least up until July by the bark reversal treatment.

The Chinese tallow tree is remarkably tolerant of poor drainage and competes exceptionally well with other species in poorly drained areas. Roots are nevertheless well developed and penetrate deeply even in heavy clay soils which are continuously submerged for 4 to 6 months of the year and subject to periods of flooding up to one month's duration during the growing season. The nature of the physiological mechanism(s) imparting tolerance to immersion is presently unknown. There is no evidence of the butt-swelling often found in some species subjected to such conditions. Preliminary observations suggest that stomata rarely close and that a dense tallow tree forest is very effective in removing excess surface water. Apparently, there is also considerable drought-tolerance, perhaps attributable to the very deep and extensive root system which allows tapping of deep supplies rather than to any innate ability to conserve water.

The Chinese tallow tree is also tolerant of a certain amount of salinity. It is commonly found along the shore-edge sand-barrier islands, along salt-creek borders, and in salt marshes. The upper limit of saline tolerance is unknown with respect to fruiting, and the mechanism(s) by which salinity tolerance is achieved are presently unknown. A prominent pair of glands in the blade-petiole junction may under certain circumstances serve as salt glands, although their only observed function thus far is in the exudation of a considerable quantity of sugar-containing fluid in the late summer and fall.

Environmental Considerations

There are significant positive and negative aspects of general environmental effects of a tree crop. First, there are climatic effects of establishing large areas of orchard or forest on land that is currently open, in low brush,

prairie, or conventional crops. Trees will tend to slow and
break the force of the wind, will add moisture to the air
while generally lowering the water table, and will change
the vertical temperature profiles by shading. There will
be a rather profound effect upon the microclimate which
when spread over sufficient area could become a modification
of the overall climate.

There are other, more complex, interactions. Monocul-
tures typically reinforce deleterious concentrations of
insect and other pests. While the Chinese tallow tree in
the United States appears not to possess a complement of
serious pests, it has probably not been naturalized in the
United States long enough for such associations to develop.
Moreover, the tree has for the most part been present in
mixture with other species and it is only in relatively
recent times that spread of the tree has reached such proportions that large brakes and forests of predominantly
Chinese tallow tree have developed. Even now it is possible
to recognize a number of insects which are associated in
specific ways with the tree. While no recognizably harmful
effects can be assigned now to the presence of these insects,
such effects could appear in time; they might be mitigated
by utilizing Chinese tallow tree as one component of inte-
grated agroforestry schemes.

The high growth rate and high levels of various phenolics in the leaves and sap of the Chinese tallow tree have
generally acted to discourage insect attack. The major
insect associations have thus been related to the very high
rate of nectar production which occurs in glands at the
base of the leaves as well as on the flowers. Recently,
however, we have observed defoliation of Chinese tallow
trees in South Texas by leaf-cutting attine ants. These
ants utilize the leaves in the culture of a fungus garden
which in turn serves as their major food source. The usual
garden fungi are higher basidiomycetes, most of which
possess the capability for detoxifying and degrading plant
phenolics. The ordinarily toxic nature of the leaf is thus
circumvented. At this point little is known about the
extent of the attine ant defoliation of Chinese tallow tree.
It is possible that the problem could be serious in tropical America where attine ants are prevalent and serious
pests.

Large numbers of tallow seeds are ingested and spread
widely by birds. If the tree were developed into a
valuable crop, the loss to birds could become an economic
concern, and the attraction and sheltering of large numbers

of birds would become a public health consideration. Techniques for dealing with birds will need to be developed.

We have pointed out that perennial crops in general reduce runoff and soil loss. But even with perennials there is always some runoff, and consideration must be given to its qualities. The Chinese tallow tree produces in its leaves a considerable amount of tannin or tannin-like materials. It has been demonstrated that reducer organisms do not attack Chinese tallow leaf until such materials have been leached out of the leaf (21). However, the fate of the leachate in the soil and water is unknown. It is known that organisms such as mosquito larvae are not present in forest pools of water containing leaf leachate (21). The effects of such metabolic products upon higher organisms are currently unknown but possibly detrimental. Presumably, a certain amount finds its way into runoff water, into estuaries and eventually into the fishery. If there are large numbers of trees present there should be correspondingly large amounts of leachate with potentially harmful effects upon various members of the food chain. Attention will have to be given to methods for isolating and/or treating such leachates.

In addition, the pattern of spread of the Chinese tallow tree must be studied. The Chinese tallow tree has the potential for becoming an extremely aggressive weed. To contemplate introducing and planting the tree over large areas without understanding its capabilities for becoming a nuisance is unwise. Experience in Texas has adequately demonstrated the ability of the tree to invade useful cropland and, once there, to maintain itself tenaciously. From field observation a number of mechanisms for spread and establishment are evident, and a general scenario can be given for spread of the Chinese tallow tree, as follows. Humans introduce a few individuals for one or more reasons; a hope of establishing a new crop, a wish to have colorful ornamental trees, or simply because the white seeds are quite attractive and it is a human habit to pick seeds of unusual appearance to plant them. These individual trees mature in 3 to 5 years and produce seeds which are sometimes picked by humans but more often are eaten by birds. Secondary spread occurs to fence rows and similar areas where birds congregate. In one area of Texas, large open-grown trees are found to be ringed in Chinese tallow tree seedlings growing beneath the periphery of the crown where birds had apparently roosted at night. As the second generation of trees matures, relatively large numbers of seeds become available for consumption by birds and are subsequently distributed in dense populations over large areas.

Disturbed land such as fallow fields, wet areas, ditches and roadsides begin to be taken over by Chinese tallow trees in populations ranging from sparse, open stands to dense brakes. In a final stage, the density of seeds becomes so great that trees begin to appear everywhere and established stands begin to encroach upon uncultivated land at the rate of several feet per year due to the movement of roots into the area. Clearly, methods for coping with unwanted spread must be developed and tested.

Wood Culture

In recent years the concept of short-rotation intensive silviculture has received considerable attention in the United States. The basic idea is that trees should be grown in much the same way as any other field crop and that the primary goal is production of as much wood material per unit area as possible, regardless of its form, with cultural inputs roughly equivalent to those for conventional annual crops. Under such conditions the expected yield would be markedly higher than yields from what is now thought of as conventional forestry.

Central to the general concept is the assumption that the plantation, once established, will be cultivated under a system of forestry practice known as coppice culture. Under this system, stumps are allowed to sprout following harvest and sprouts are then allowed to grow to harvestable age. This cycle is repeated through a number of harvests. The advantages of coppice over conventional forestry include: (a) greater yield, (b) lower average cultural costs, because growth of the tree is not hindered by the necessity for establishment of an extensive root system and cost of establishment is a one-time cost with harvest costs being the major costs in subsequent rotations, and (c) lower average impact upon the environment, again because the soil disturbance associated with land preparation and planting only occurs once in several cropping cycles. Such woody material might form a low-cost resource base for industrial use as fuel, fiber or chemical feedstock.

The above concept has undergone a considerable amount of change in the past few years as several economic realities have become apparent. Prices of land have risen sharply, cultural costs have increased with increase in the cost of labor and fuel and most importantly, the majority of the species considered useful for short-rotation culture must be planted as seedlings or cuttings. Production of such materials as well as planting them is relatively labor intensive. Cost of stand establishment at the densities

necessary for high yield on short rotations is thus very
high and at high interest rates, is an investment that is
difficult to recover from a relatively low-value wood crop.
Nor does the fact that average cultural costs are substantially lowered by the use of the coppice system improve the
economics sufficiently to compensate for high interest
costs.

Four years of research at the University of Houston by
two of us (JRC, HWS) have yielded relatively positive indications that the Chinese tallow tree may be cultivated in
such a way as to eliminate the above economic objections (2).
Although the wood is relatively weak and short-fibered, and
stems adequate as sawlogs are seldom produced, the caloric
content is adequate (7500 Btu/lb). When properly dried,
the wood burns well as a domestic heating or cooking fuel.
Productivity of wood mass is high; four year plantations in
the Houston area produced in excess of 4 tons of oven dry
wood/acre/year with relatively minor cultural inputs.

Seeds, Tallow, and Kernel Oil

The seed of the Chinese tallow tree is comprised of an
outer coating of hard tallow embedded in a fibrous matrix
separated from an oily seed kernel by a hard, impermeable
shell. There are two triglyceride products: solid, relatively saturated fat, called "tallow", and relatively
unsaturated liquid kernel oil. In addition, there is a
high-protein meal. One of us (RK) has analyzed a number of
Chinese tallow tree seed samples for tallow and oil (Table 1).
The seed meal, amounting to roughly 10 percent of the seed
weight, contains about 70 percent of a protein that is
somewhat deficient in lysine and methionine (3). Properties
of the seed shell, about 40 percent of the seed weight, are
presently unknown.

Of special interest is the "estolide" (tetraester triglyceride) fraction of the kernel oil (Figure 2) containing
two unusual short-chain fatty acids, 8-hydroxy-5,6-octadie-
noic acid (HODA), and trans-2,cis-4-decadienoic acid (DDA).
Industrial uses are yet to be defined.

Seed yields in excess of 10,000 lbs/acre have been
estimated for the Chinese tallow tree (14) based on reported
yields from plantations of the original U.S. Bureau of Plant
Industry trials carried out near Houston, Texas and Jacksonville, Florida in the early part of this century. All data
are from secondary references -- the original reports are
no longer available -- but the yields reportedly would be
achieved in orchards having trees spaced at 16.5 feet or

Table 1. Variability of seed tallow and oil in Chinese tallow tree seeds collected in the Houston, Texas area.

Tree Number	% Lipid			Mole % Estolide*	Oil Composition %			Tallow Composition %	
	Oil	Tallow	Total		18:1	18:2	18:3	16:0	18:1
1	23.2	14.4	37.6	28	13.3	29.2	48.8	[76.4]	(21.7)
5	19.1	26.2	45.3	[37]	13.1	34.3	44.0	69.6	28.3
8	[25.1]	20.3	45.4	31	12.9	26.9	51.1	71.0	26.9
12	22.1	(13.1)	35.2	(27)	13.1	34.7	44.0	67.5	29.2
14	15.3	[36.4]	[51.7]	27	19.8	27.9	41.2	67.7	29.7
18	18.4	26.0	44.4	28	(11.1)	26.7	[54.9]	68.9	28.7
20	20.8	21.6	42.4	31	15.4	(25.8)	50.2	67.6	29.3
21	22.1	18.6	40.7	32	13.8	[37.6]	(40.8)	66.3	30.8
31	19.1	28.5	47.6	31	13.4	32.5	45.4	(65.8)	[31.9]
38	(14.0)	20.0	(34.0)	27	14.6	28.3	48.7	68.1	30.1
44	18.9	30.6	49.5	29	[22.1]	24.2	44.4	71.1	27.3

* Mole percent of estolide-containing glycerides in the oil fraction.

Notes: Samples of seeds were collected from fifty trees selected for high yield and large seed size. Tallow was extracted from intact seeds with hot hexane. Seeds were crushed and oil was recovered by re-extraction. Fatty acids were quantitated by gas chromotography. Estolides were quantitated by HPLC, utilizing infrared detection. Analytical results are only reported here for trees having the highest [brackets] or lowest (parentheses) values for each component.

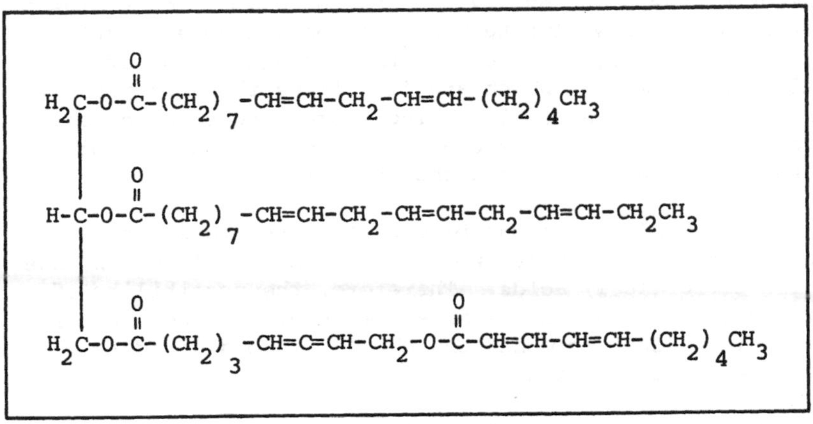

$$\text{H}_2\text{C-O-}\overset{\displaystyle O}{\overset{\|}{\text{C}}}\text{-(CH}_2)_7\text{-CH=CH-CH}_2\text{-CH=CH-(CH}_2)_4\text{CH}_3$$

$$\text{H-C-O-}\overset{\displaystyle O}{\overset{\|}{\text{C}}}\text{-(CH}_2)_7\text{-CH=CH-CH}_2\text{-CH=CH-CH}_2\text{-CH=CH-CH}_2\text{CH}_3$$

$$\text{H}_2\text{C-O-}\overset{\displaystyle O}{\overset{\|}{\text{C}}}\text{-(CH}_2)_3\text{-CH=C=CH-CH}_2\text{-O-}\overset{\displaystyle O}{\overset{\|}{\text{C}}}\text{-CH=CH-CH=CH-(CH}_2)_4\text{CH}_3$$

Figure 2. Typical estolide (tetraester triglyceride) molecule found in the kernel oil of the Chinese tallow tree.

160 trees/acre in which the average yield of each tree was
70 lbs of seeds. From this one may calculate a yield of
11,200 lbs of seeds per acre. A few percent of moisture
may reasonably be expected, and so a round figure of 10,000
lbs of seed per acre is probably to be expected. These
yields are much higher than those of annual oilseed crops.
For example, peanuts in the United States yield about 700
lbs of oil per acre, the highest per acre oil yield of all
United States oilseeds (22). Chinese tallow tree yields of
total fat (tallow plus kernel oil), estimated at about
4500 lb per acre, would exceed the average yields of
Malaysian palm oil, which are five times those of peanut,
or about 3500 lb/acre (22). Therefore, the Chinese tallow
tree must be regarded as one of the most productive oilseeds of the world, if not the most productive.

From Table 1, it can be seen that the tallow fatty
acids are approximately two-thirds palmitic (16:0) and one-
third oleic (18:1) acids. Therefore, the possibility of
utilization of the tallow as a component of cocoa butter
equivalent (CBE) formulations should be considered as well
as other food fat uses.

The highly unsaturated nature of Chinese tallow tree
kernel oil suggests its use as a drying oil. Tests by
Bolley and McCormack compared Chinese tallow tree kernel
oil, bleached linseed oil, and unbodied dehydrated castor
oil in paints and varnishes (15). All three oils gave
dried films with good resistance to cold and hot water and
alkali, and the Chinese tallow tree kernel oil film was the
most insoluble of the three. Furthermore, a varnish made
from it was harder than varnishes made from linseed or
castor oil. Bolley and McCormack concluded that the kernel
oil:

> "is an excellent general purpose drying oil for use
> in paints and varnishes and would rate for many
> purposes superior to bleached linseed oil, while
> approximating the desirable characteristics of
> dehydrated castor oil" (15).

The two unusual short-chain fatty acids available from
the estolide fraction of the kernel oil might be used in
novel polymers or other fatty acid-based chemical products.
For example, the allenic system in the 8-carbon HODA provides the synthetic organic chemist with an unusually long
1,3-disubstituted allene. The difunctional character of
HODA suggests the possibility of allenic polyesters or
polyamides. A variety of polymers might be possible from
DDA by way of Diels-Alder reactions. DDA might be converted

to dibasic acids useful as unusual polymer precursors, or as esters for fat-based synthetic lubricants. Also of interest is the fact that the methyl ester of HODA strongly inhibits such microorganisms as <u>Streptococcus faecalis</u> and <u>Pseudomonas aeruginosa</u> (23). Further studies on the biocidal activity of HODA esters should be carried out.

Other Products

Other products which may contribute to overall economic viability are leaf extractives and honey. A black dye for silk was extracted from the leaves by the Chinese (4, 10, 12). Although the properties of this material are unreported it is possible that Chinese tallow tree leaf could serve as a source of industrial tannins. The Chinese tallow tree provides a major honey crop in the areas where it is prevalent (24). In preliminary experiments at the University of Houston Coastal Center, 400 bee colonies produced a remarkable 30,000 lbs of honey with a wholesale value of $15,000 from approximately 300 acres of Chinese tallow tree forest.

Economics of Crop Growth

Assuming rough equivalence of tallow, kernel oil and seed meal to palm oil, linseed oil and linseed meal, respectively, and making further assumptions concerning transportation, storage, conversion and processor's profits, a net revenue of about $1300 acre/year from a mature Chinese tallow tree orchard has been estimated (1). This is attractive compared to the equivalent revenue for most field crops, although such a yield would not be expected until maturity, no earlier than five years after planting. Stand establishment by direct planting of seed appears to be a relatively inexpensive operation, well suited for use with the Chinese tallow tree in short-rotation silviculture (2). However, it is now evident that seed orchards must be established from carefully selected clonal materials to obtain the uniformity of growth pattern and seed ripening necessary for use with large-scale mechanized cultivation. Barring breakthroughs in use of growth regulators or in propagation technology, orchard establishment is likely to be expensive. It seems likely that economic success of the Chinese tallow tree in the United States will depend upon such breakthroughs.

Genetics and Reproductive Strategy

The Chinese tallow tree exhibits a remarkable degree of variability, presenting a nearly bewildering variety of traits for selection. For example, Table 1 illustrates the great variability in seed chemistry available in seeds from fifty trees picked virtually at random from a population of many millions in one locality. Such chemical variability indicates the possibility of tailoring varieties of trees for specific products and, taken with the observed differences in growth habit, promises the ability to adapt to a wide variety of growth situations as well.

The reproductive strategy is of interest for two reasons: First, it has considerable influence upon the genetics and thus upon the way future crop improvement programs are planned. Second, an understanding of the way flowering occurs is important to an understanding of the way the crop of interest is formed.

Experiments during the past three years have shown that flowering and seed formation in the Chinese tallow tree is somewhat complex. There are two basically different types of seed clusters; "grape," and "eagle claw." Normally, a given tree will have only one type. While both types produce male and female flowers, the times of ripening of male and female flowers on a given tree are different (25). The males of the "grape" type tend to produce pollen at the time the females of the "eagle claw" type are receptive and vice versa. Thus, this out-pollination mechanism contributes to the high variability, and at a more practical level dictates that plantations must contain trees of both types. Moreover, the pollenation at different times causes differences in maturity and ripening of seeds.

Agronomic Practices

In the United States a Chinese tallow tree cultivation system is most appropriate for large areas of generally less desirable or agriculturally marginal land, and cultivation and harvest must be highly mechanized. The traditional Chinese culture of Chinese tallow tree has utilized an orchard configuration in which trees are planted in regular evenly spaced array with sufficient space allowed for full crown development and with pruning of the crown to aid in hand harvesting. Such practices allow considerable flexibility of method and time of harvest. Under the kinds of cropping systems that appear mandatory for agriculture in the United States, such flexibility is not possible. The requirement for mechanical harvest dictates the use of some uniform plant configuration, probably of a hedgerow type (26) and of large dimensions, thus requiring relatively large equipment. The natural consequence of such a system of cultivation will be a need for very uniform plant

material in terms of form, ripening time, leaf fall, etc.,
requiring a substantial development effort to bring all
elements to maturity.

Seed Handling and Processing

Because Chinese tallow tree seeds are similar in shape
and size to soybeans, the physical facilities for soybean
transport and storage should be usable. The major unan-
swered question concerns degradation of the outer tallow
coating. There is some observational evidence which indicates a tendency toward oxidation during storage. Quantita-
tion of the observed effect and development of preventative
measures will be necessary.

The seed produces two separate fat products; therefore,
processing methods will be somewhat complex. The traditional Chinese methods involve steaming or hot water treatment and abrasion to remove the tallow from intact seeds,
followed by crushing and pressing to recover the liquid oil
from the kernel. One of us (CRE) has shown that the same
result may be accomplished using solvent extraction and
standard modern processing equipment (Figure 2). Whole seeds
were extracted with hexane in a pilot batch extraction
apparatus, dried and decorticated using a Carver bar huller.
Screening separated large hull pieces from small pieces and
endosperm which were re-extracted with hexane. Oil and
tallow were recovered from the solvent by vacuum distillation.

Potential for Developing Countries

It would seem feasible to adapt traditional labor-
intensive Chinese methods of cultivation and processing to
many developing countries, considering first those that are
semi-tropical, with the possibility of extension to certain
tropical areas, especially mountainous tropics (27). The
tree is found in several South and Southeast Asian countries,
as well as in China below 32° north (36° along the coast).
Fuel uses of the seeds, tallow, and oil, oriented to the
needs of rural people in developing nations, have been
demonstrated by Mathieu (28), including the burning of whole
and crushed seeds to cook meals, and the burning of the
crushed seeds and the expressed kernel oil for lighting. A
description of these tests is included in Chapter 12. The
seeds are an attractive fuel, with a heating value of
12,000 $\pm$ 600 Btu/lb. (29). Potential non-fuel uses of
Chinese tallow tree seeds in developing countries include
for the tallow: food fat, cocoa butter extender, and candle

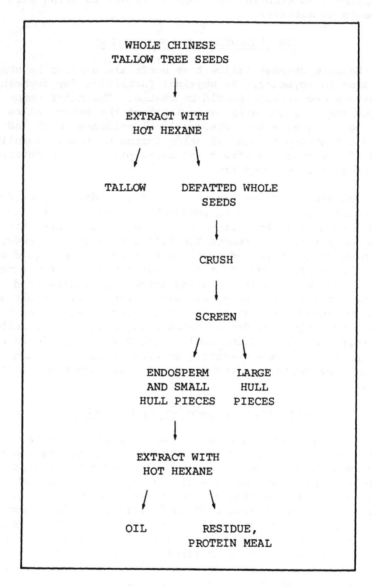

Figure 3. Processing flow chart for double solvent extraction of Chinese tallow tree seeds.

manufacture; for the kernel oil: drying oil and chemical
intermediate uses (27).

Acknowledgements

The authors are grateful for support provided by a U.S.
Department of Energy grant to the University of Houston.
Additional support was given by the National Science Foundation Small Business Innovation Program and the United States
Department of Agriculture through grants to Simco, Incorporated. Studies of product utilization at The Center for
Development Technology, Washington University, were funded
by Simco, Inc., under the U.S. Department of Agriculture
grant. Seed analyses were performed by the Northern Regional
Research Center of the USDA, and studies of the extraction
of tallow and kernel oil were carried out by the Food Protein Research Center of Texas A&M University. Any opinions,
findings and conclusions or recommendations expressed herein
are those of the authors and do not necessarily reflect the
views of the sponsoring organizations.

References

1. H. W. Scheld, N. B. Bell, G. N. Cameron, J. R. Cowles,
    C. R. Engler, A. D. Krikorian and E. B. Shultz, Jr.
    "The Chinese Tallow Tree as a Cash and Petroleum-
    Substitute Crop," In Tree Crops for Energy Co-Production
    on Farms. SERI Bull. CP-622-1086, (1981) 97-111.
2. H. W. Scheld and J. R. Cowles. "Woody Biomass Poten-
    tial of the Chinese Tallow Tree," Econ. Bot., 35 (1981)
    391-397.
3. B. R. Holland and W. W. Meinke. "Chinese Tallow Nut
    Protein. I. Isolation, Amino Acid, and Vitamin Analy-
    sis," J. Am. Oil Chem. Soc. 25 (1948) 418-419.
4. J. D. MacGowan. "Uses of Stillingia sebifera or Tallow
    Tree With a Notice of the Pe-La, an Insect-Wax of
    China," Am. J. Sci. 12 (1851) 17-22.
5. Anonymous. "Inventory of Seeds and Plants Imported by
    the Office of Foreign Seed and Plant Introduction During
    the Period from April 1 to May 31, 1920." U.S.D.A.
    Bureau Plant Indust. Inventory No. 63 - Nos. 49797 to
    50646 (1923).
6. C-L. Shih. (Untitled. Deals, in Chinese, with the
    breeding and culture of the Chinese tallow tree Sapium
    sebiferum Roxb.). Research Bulletin of the Chekiang
    For. Res. Inst., People's Republic of China, 1973.
7. R. Kellison. Personal communication. (March 1982).
8. S-K. Lee. "Genus Sapium in the Chinese Flora," Acta
    Phytotaxonomica 5 (1956) 111-130.

9. K. A. Williams. "Oils, Fats, and Fatty Foods." Amer. Elsevier Publ. Co., Inc.: New York. (1966).

10. W-C. Lin, A-C. Chen, C-J. Tseng and S-G. Huang. "Chinese Tallow Tree in Taiwan (Sapium sebiferum)," Bull. Taiwan For. Res. Inst. 47 (1958) 1-37.

11. R. Hurov. Personal communication. (January 1980).

12. F. N. Howes. "The Chinese Tallow Tree (Sapium sebiferum Roxb.) -- A Source of Drying Oil." Kew Bull. 4 (1949) 573-580.

13. G. S. Jamieson and R. S. McKinney. "Stillingia Oil." Oil and Soap 15 (1938) 295-296.

14. W. M. Potts. "The Chinese Tallow Tree as a Chemurgic Crop," The Chemurgic Digest 5 (1946) 373-375.

15. N. S. Bolley and R. H. McCormack. "Utilization of the Seed of the Chinese Tallow Tree," J. Am. Oil Chem. Soc. 27 (1950) 84-87.

16. F. W. Kahn, K. Khan and M. N. Malik. "Vegetable Tallow and Stillingia Oil from the Fruits of Sapium sebiferum Roxb." Pakistan J. of Forestry 23 (1973) 257-263.

17. M. U. Zubair and Z. Zaheer. "New Possible Indigenous Source of Fat," Pakistan J. Sci. Ind. Res. 21 (1978) 136-137.

18. J. Teas. Personal communication. (September 1979).

19. J. Hutchison. Personal communication. (June 1979).

20. M. Bell III. "Some Notes and Reflections Upon a Letter from Benjamin Franklin to Nobel Wimberley Jones, October 7, 1772", Ashantilly Press, Darian, Georgia (1966).

21. G. N. Cameron and T. W. LaPoint. "Effects of Tannins on the Decomposition of Chinese Tallow Leaves by Terrestrial and Aquatic Invertebrates," Oecologia 32 (1978) 349-366.

22. H. O. Doty, Jr. "Economics of Oilseed Production," Presented at the 23rd Annual Meeting of the Society for Economic Botany, Tuscaloosa, AL., June 14-16, 1982.

23. H. Ohigashi, K. Kawazu, H. Egawa, and T. Mitsui, "Antifungal Constituent of Sapium japonicum," Agr. Biol. Chem. 36 (1972) 1399-1403.

24. B. Hayes. "The Chinese Tallow Tree (Sapium sebiferum) -- Artificial Bee Pasturage Success Story," Am. Bee J. 119 (1979) 848-849.

25. M. Vrecenar, J. R. Cowles and H. W. Scheld. "Patterns of Flower Development in the Chinese Tallow Tree (Sapium sebiferum (L.) Roxb.)." In preparation (1982).

26. S. I. Mason, Jr. and H. W. Scheld, U.S. Patent 4,327,521, May 4, 1982.

27. H. M. Draper, III. "New Oilseed Crops for Fuels and Chemicals: Ecological and Agricultural Considerations," D.Sc. Dissertation, School of Engineering and Applied Science, Washington University, St. Louis, MO, (December, 1982).

28. S. L. Mathieu. "Potential Utilization of Oilseeds for Household Energy at the Village Level," M.S. Thesis, Department of Technology and Human Affairs, Washington University, St. Louis, MO, (May, 1982).

29. H. W. Scheld. Unpublished. Based on measurements carried out by the University of Texas (Austin) Department of Chemistry, (1980).

Anna-Maria V. Watowich, Eugene B. Shultz, Jr.

7. Polymers from Novel Oilseeds in the Southeastern United States

<u>Introduction</u>

Because the polymer industry is heavily dependent on nonrenewable petroleum-based feedstocks, there may be value in exploring alternatives such as domestic U.S. renewable resources of fatty acids from seed oils. Unusual fatty acids such as erucic, eicosenoic and lesquerolic acids can be derived from the seeds of a number of species of unconventional oilseed plants that might be cultivated on marginal lands. Certain engineering thermoplastics, nylons-11, -12, 13 and -13/13 as well as others, might be manufactured from these sources competitively, and the by-products utilized in the synthetic lubricant and other industries.

This concept differs in at least two ways from some other approaches to alternative sources of chemical feedstocks (1) that involve coal, oil shale, or alcohol. By contrast:

a) Our orientation is toward small-volume, high-unit-price specialty chemical production from renewable resources, rather than large-volume, low-unit-price commodity chemicals based on low-molecular-weight feedstocks.

b) We are interested in utilizing long-chain molecules with unusual structural features that can be found in vegetable oils extracted from certain seeds, rather than finding new sources of the small, simple molecules typical of current petrochemical starting materials.

Oilseeds, Seed Oils, Triglycerides
and Fatty Acids

There are thousands of plants known to produce significant amounts of fats and oils in their seeds (2). However,
it has only been in the past twenty years that the chemical
compositions of the fats and oils of many of these plants
have been investigated. Much of this work on the analysis
of oil-rich seed species has been undertaken by the U.S.
Department of Agriculture's Northern Regional Research
Center in Peoria, Illinois. Since 1958, investigators at
this institution have studied 6500 species of wild plants
(3). Although there are thousands of species with oil-rich
seeds, only a handful are commercially grown, such as sunflower, peanut and soybean. Examples of some unusual oilseeds with possible application to the southeastern United
States are listed in Table 1. Some of these have been
discussed in Chapter 2.

Table 1. Some unusual oilseeds adapted to the southeastern
United States, and of potential interest as renewable
resources of polymers

Species	Type	Fatty Acid	Polymer Products
Crambe abyssinica (crambe)	annual	erucic	higher nylons
Eruca sativa (rocket salad)	annual	erucic	higher nylons
Lesquerella globosa (lesquerella)	perennial	lesquerolic	higher nylons
Marshallia caespitosa (Barbara's buttons)	perennial	eicosenoic	higher nylons
Stokesia laevis (Stoke's aster)	perennial	vernolic	epoxies

The oil content in a seed can range from a low of a
few percent in some species, to a high of 70 to 80 percent
in other species (4). In about 6000 of the seeds analyzed
by the USDA research group in Peoria, 42 percent contained
more than 20 percent oil, by weight (5). The oils derived

from such oilseeds are usually glycerides composed of glycerol esterified with one, two or three long-chain carboxylic acids. The carboxylic acids usually have an even number of carbon atoms per molecule, in the range between 8 and 24. Because they originated from fats or oils, these acids are referred to as fatty acids (6). Three fatty acids esterified to one glycerol molecule is a triglyceride, and triglycerides are usually the most commonly found glycerides in plant oils.

The oil in a seed can be extracted from the seed by hydraulic presses, expellers (screw presses), or by solvent extraction. The crude oil extracted from the seed must be further refined to remove suspended solids, color, odor or free fatty acids. Settling, degumming and acid washings will often remove the suspended solids. Bleaching and deodorizing techniques are used to remove undesirable colors and odors. Alkali treatment removes free fatty acids. After refining, the oil can be hydrolyzed to obtain the components of the triglycerides, namely glycerol and the fatty acids.

Fatty acids are long straight-chain carboxylic acids, and may be saturated or unsaturated. Certain fatty acids have hydroxy groups, epoxy groups, allenic functions and cis- and trans- unsaturation. Table 2 provides a list of some unusual fatty acids and their chemical structures. The occurrence of such fatty acids with a variety of functional groups increases the number of possible chemical reactions and end-products.

In 1980, U.S. production of fatty acids via hydrolysis of seed oil triglycerides was over one billion pounds (7). About one-third is used for surfactants, 18 percent for fatty nitrogen products, 10 percent for rubber products, 10 percent for protective coatings, and lesser percentages for grease products, textile industry products, plasticizers, food additives, cosmetics, and pharmaceuticals (8). At present, fats and oils participate in the engineering thermoplastics market only to the extent of about 2 percent (9). Additional information on markets for fats and oils is given in Chapter 4.

This chapter deals mainly with specialty nylons that might be derived from unusual fatty acids obtainable from southeastern U.S. seed oils that are presently noncommercial. Nylons may be formed by the condensation polymerization of a dicarboxylic acid to a diamine, or by self-condensation of an amino carboxylic acid.

Table 2. Chemical structures of some unusual fatty acids

Acid	Source	Structure
Erucic	Crambe abyssinica Eruca sativa	$CH_3 (CH_2)_7 - CH = CH - (CH_2)_{11} - COOH$
Eicosenoic	Marshallia caespitosa	$CH_3 (CH_2)_7 - CH = CH (CH_2)_9 - COOH$
Petroselinic	Foeniculum vulgare	$CH_3 (CH_2)_{10} - CH = CH - (CH_2)_4 - COOH$
Ricinoleic	Ricinus communis	$CH_3 (CH_2)_5 - \underset{\underset{OH}{\vert}}{CH} - CH_2 - CH = CH - (CH_2)_7 - COOH$
Lesquerolic	Lesquerella globosa	$CH_3 (CH_2)_5 - \underset{\underset{OH}{\vert}}{CH} - CH_2 - CH = CH - (CH_2)_9 - COOH$
Vernolic	Stokesia laevis	$CH_3 (CH_2)_4 - \overset{\displaystyle O}{\overbrace{CH - CH}} - CH_2 - CH = CH - (CH_2)_7 - COOH$

Commercial Nylon Polymers

The commercial nylons produced in greatest quantities are nylon-6 and nylon-6/6. These have been used to make textile and carpet fibers, brush filaments, and plastic film. Nylon is injection molded for appliances, gear wheels, bearings and other small mechanical parts, in addition to being used for wire and cable coverings (10). Other nylons commercially available include nylon -6/10, -6/12, -11 and -12. The production of all nylons totalled 316.8 million pounds in 1979. In 1980, the cost of nylons ranged from $1.64 per pound for nylon-6 and -6/6 to $2.57 per pound (11) for nylon-11. It is such higher-priced nylons as nylon-11, -12, -13, and -13/13 that are of greatest interest to us. The higher numbers (nylons 11 through 13) indicate longer hydrocarbon chains than are found in nylon -6 and -6/6, leading to the name "higher nylons". Because of the longer hydrocarbon chains, higher nylons have special properties that often allow them to command higher unit prices, in comparison with 6-carbon nylons.

For example, nylon-11 differs significantly from the conventional nylons (-6 and 6/6) in these ways (12):

1) It has a relatively low melting temperature, making the plastic easier to process and less heat degradable.

2) It has lower moisture absorption making the plastic dimensionally stable, as well as capable of retaining its electrical properties under a variety of conditions.

3) It has good low-temperature properties.

4) It has greater flexibility (high modulus of elasticity).

5) It has good resistance to a broad range of chemicals, allowing it to be used as a protective coating on many metals.

Nylon-11, which has properties very similar to those of nylon-12, is the only commercially available nylon produced from vegetable oil. It is a specialty product formed by polymerization of 11-aminoundecanoic acid which is a derivative of ricinoleic acid, the main fatty acid of castor oil. Figure 1 shows how nylon-11 is produced from ricinoleic acid.

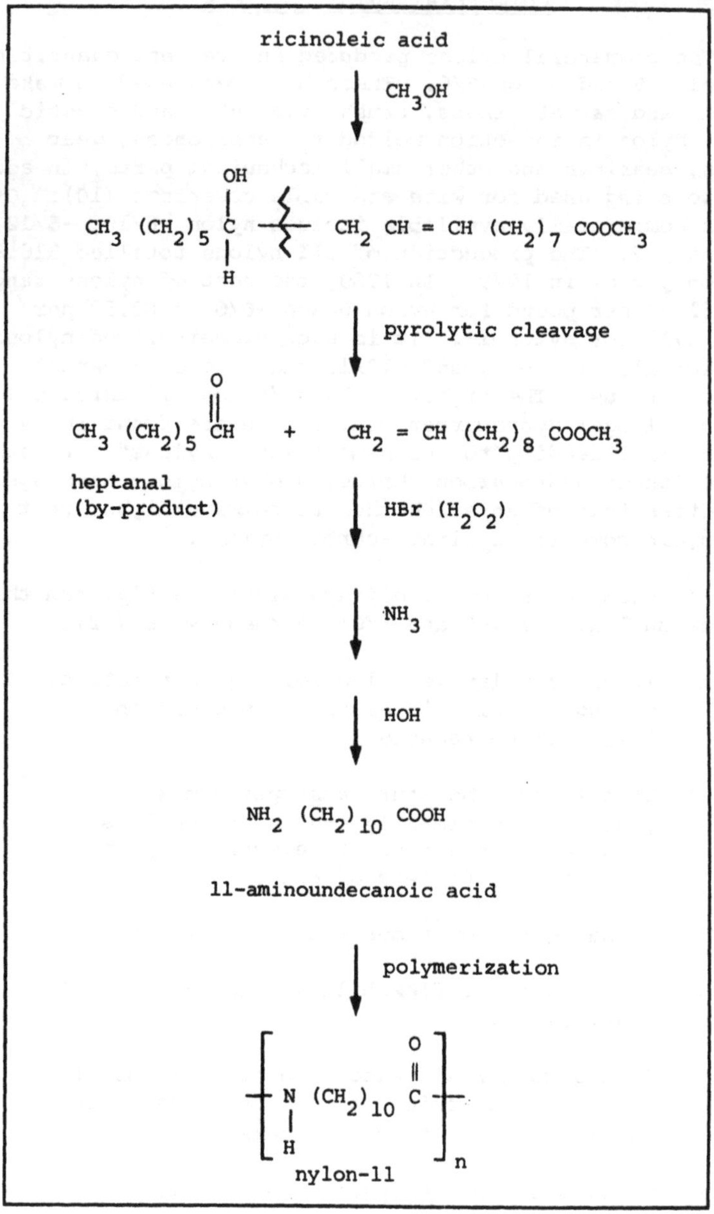

ricinoleic acid

$\downarrow$ CH_3OH

$$CH_3 \ (CH_2)_5 \ \underset{\underset{H}{|}}{\overset{\overset{OH}{|}}{C}} \ \{ \ \ CH_2 \ CH = CH \ (CH_2)_7 \ COOCH_3$$

$\downarrow$ pyrolytic cleavage

$$CH_3 \ (CH_2)_5 \ \overset{\overset{O}{\parallel}}{CH} \quad + \quad CH_2 = CH \ (CH_2)_8 \ COOCH_3$$

heptanal
(by-product)

$\downarrow$ HBr (H_2O_2)

$\downarrow$ NH_3

$\downarrow$ HOH

$$NH_2 \ (CH_2)_{10} \ COOH$$

11-aminoundecanoic acid

$\downarrow$ polymerization

$$\left[\ \overset{}{\underset{\underset{H}{|}}{N}} \ (CH_2)_{10} \ \overset{\overset{O}{\parallel}}{C} \ \right]_n$$

nylon-11

Figure 1.   Commercial Production of Nylon-11 from Ricinoleic
Acid.   (Source: Reference 12).

The sole producer of nylon-11 is ATO Chimie, a French chemical company. The monomer is produced in France and then polymerized in the United States by Rilsan Corporation, a subsidary of ATO, in Birdsboro, Pennsylvania. Through the use of additives, Rilsan makes over 80 grades of nylon-11 plastics covering a broad range of properties (12). In addition, Rilsan supplies other companies with the polymer to produce their own plastics grades. ATO Chimie developed and commercially introduced nylon-11 to the European marketplace in 1955 (13). It was not until manufacturing facilities were put on stream in Pennsylvania in 1971 that Rilsan Corporation marketed nylon-11 in the U.S.

Nylon-11 can be injection molded, rotation molded, blow molded or extruded into a variety of products, or it can be used in a powder form to produce a protective coating on metal surfaces. The list of products that are made from this specialty nylon is substantial. Fishing lines, soccer shoe soles, ski boot skells, air brakehoses and molded gears are a few of the many products made with nylon-11.

Some of the established applications for nylon-11 in Europe that currently remain rather limited in the U.S. include (14):

1) Fuel line material.

2) Food packaging film.

3) Metal coating for hospital furniture, outdoor furniture and mechanical components.

Since 1971, many specialty applications for nylon-11 have been found and exploited in the U.S. An increase in capacity by 65 percent at the Birdsboro plant indicates that there is a demand for nylon-11. Given that nylon-11 is now well-accepted in the U.S. polymer marketplace, it may be interesting to investigate what other nylons may exist that could compete with the castor oil-derived nylon-11, and the petroleum-derived similar plastic, nylon-12, the only two higher nylons commercially available. If petroleum continues to escalate in price over the long term, vegetable oil sources may become more attractive. Further, since the processing of castor beans involves allergenic hazards to health (15), it is not likely that castor will be grown in the United States; it is presently grown in Third World countries for ATO Chimie.

Higher Nylons Not Yet Commercial:
The Nylon-13s

The nylon-13s (-13, -13/13 and -6/13) have been studied in depth at the USDA Northern Regional Research Center in Peoria, Illinois. Most of the work there on the nylon-13s was done during the 1960s and ended in the early 1970s. Erucic acid was studied as a likely source for the 13-carbon brassylic acid, a monomer for nylon-13/13 and -6/13. Erucic acid is the most important constituent of the seed oils of Crambe abyssinica (Crambe), Eruca sativa (rocket salad) and Iberis amara (rocket candytuft). All are annuals potentially applicable to ecological zones of the Southeast (16). Erucic acid is a 22-carbon fatty acid with a double bond at carbon-13 (see Table 2).

The current commercial source of erucic acid is from high-erucic rapeseed. Because rapeseed is not grown domestically, rapeseed oil must be imported from Canada. In the late 1960s Canadian researchers developed strains of rapeseed having low-erucic acid content (17). This low-erucic rapeseed has since been planted in Canada to provide an edible oil and feed meal. In 1976, the U.S. experienced a shortage of erucamide, a plastics additive derived from erucic acid, because high-erucic acid rapeseed oils had been supplanted by lower-erucic acid rapeseed oils for edible purposes (18). The demand for high-erucic rapeseed oil for industrial applications along with the dependence upon foreign suppliers who may be supplanting high-erucic rapeseed with low-erucic rapeseed has resulted in research being undertaken to find alternative high-erucic acid crops (19).

Crambe, an annual oilseed crop related to the rapeseed and mustard family (family Cruciferae), has been considered in detail as a new oilseed crop by the Northern Regional Research Center. Seeds contain 30 to 40 percent oil of which 55 to 60 percent is erucic acid (17). Agronomic and genetic studies have been performed on the oilseed to determine its adaptability and variability in a variety of climates (20-22). Crambe has been grown semicommercially from North Dakota to Texas, and from California to Connecticut using conventional farming equipment. Typical yields range from 1600 to 2400 pounds of seed per acre. The seed can be harvested from 90 to 100 days after planting, making two harvests in the southern U.S. a possibility (17). However, crambe seems to be better adapted to areas where its major growth can occur during a relatively cool season (21). Therefore, many southeastern states may not be appropriate. Kentucky, however, is a potential crambe-producing state;

the average yield was 1784 pounds of seed per acre (2000 kg
per hectare) in 1977 (22). Eruca sativa and Iberis amara
may be more suitable for more states of the Southeast, in
comparison with crambe, but crambe has received much more
agronomic attention.

There are several potential uses for erucic acid in
the chemical industry (19). One of the most interesting is
the production of brassylic acid which has many uses including, potentially, the making of higher nylons. Typical
reactions, demonstrated but not commercialized, are out-
lined in Figures 2 and 3. The two methods shown in Figure
2 for the production of nylon-13 were found to give the
best overall yields for making 13-aminotridecanoic acid
(monomer of nylon-13) when the original experimental work
was done in 1969. The route using eruconitrile (ammonia-
tion of erucic acid) gave pure 13-aminotridecanoic acid in
55 to 60 percent yields. The alternate method using eru-
cate (esterified erucic acid) had an overall yield of about
48 percent. Analogous reactions using eicosenoic acid,
leading to nylon-11 and -11/11, should probably be explored.
Marshallia caespitosa and Lepidium virginicum are southeastern plants with high percentages of eicosenoic acid in
their seed oils (see Chapter 2). Eicosenoic acid is a
20-carbon fatty acid with a double bond at carbon-11 (see
Table 2).

Pilot-scale production of brassylic acid and the 13-
carbon diamine, 1,13-diaminotridecane, gave yields for
brassylic acid from erucic acid ranging from 72 to 82 percent, and yields of the diamine ranging from 78 to 96 percent (24, 25). The overall yields of nylon-13/13 are
higher, and the process appears to be simpler than those of
nylon-11, -12 and -13. The initial oxidation step results
in the production of pelargonic acid which is a valuable
by-product used in synthetic lubricants, a growing market.
Brassylic acid can be combined with other diamines to give
other nylons. For example, hexamethylene diamine, a 6-
carbon diamine used to make nylon-6/6, can be combined with
brassylic acid to make nylon-6/13.

A comparison of the physical properties of nylons-13,
-13/13 and -6/13 with those of other nylons reveals that
nylons-13 and -13/13 are quite similar to nylons-11 and -12,
whereas nylon-6/13 is similar to nylon-6/10. Nylons-13 and
-13/13 exhibit three properties that are different enough
from nylons-11 and -12 to be of importance. The most significant difference is the decreased sensitivity of nylons-13
and -13/13 to water, most apparent for nylon-13/13. This
results in greater resistance to electrical current, and a

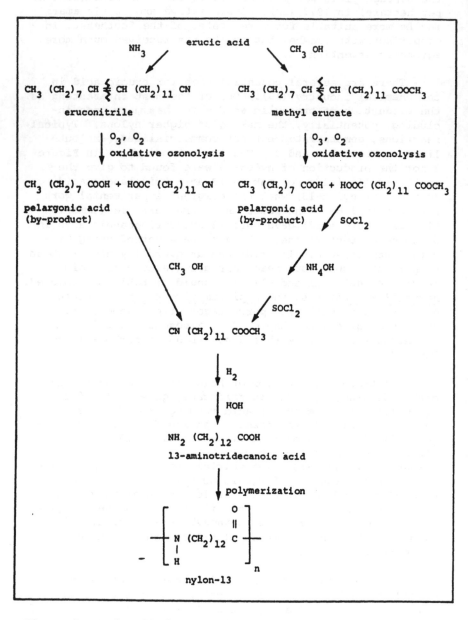

Figure 2.   Nylon-13 from erucic acid (demonstrated, but not commercialized).   (Source: Reference 23)

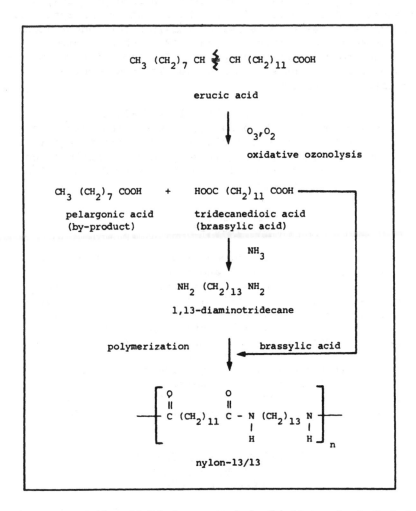

Figure 3. Nylon-13/13 from erucic acid (demonstrated, but not commercialized). (Source: References 24, 25)

dielectric strength of the nylon-13s greater than that of
nylons-11 and -12 under varying humidity conditions (26).
This property could be of importance in some products, such
as electrical cable coverings. The low water absorption of
the nylon-13s makes these nylons feasible for end-products
similar to those of nylons-11 and -12, such as tubings,
hoses and molded mechanical parts. The low moisture absorption of these nylons should also result in exceptional
dimensional stability.

Since the markets for nylons-11 and -12 are well
defined and appear to be growing, it seems reasonable to
expect that the nylon-13s could enter the marketplace if
they were available at competitive prices. Therefore, consideration of the economic feasibility of producing nylon-
13s appears to be timely.

Estimation of Production Costs
of Nylon-13/13

Sohns (27) estimated production costs in 1970 dollars
of brassylic acid (99% pure) and nylons-13, -13/13 and
-6/13. The results indicated that the nylons-13s were
economically feasible at that time. Later, the estimates
were updated to mid-1975 dollars (24, 25). The estimate
for nylon-13/13 has now been further updated to mid-1981
by Watowich (28), indicating that, with credit for the
by-product pelargonic acid, the cost was about 40 cents per
pound less than the 1981 selling price of nylon-11 ($2.57
per pound). It would appear, therefore, that with by-
product credit included in the production cost that nylon-
13/13 might have competed economically in 1981 with nylon-
11.

Southeastern Lesquerella as a
Potential Source of
Specialty Nylons

The many Lesquerella species of oilseed crops are
promising domestic sources of lesquerolic acid (Table 2),
the 20-carbon homolog of the ricinoleic acid from castor
oil. Castor oil is the main source of hydroxy fatty acids
in the U.S. and must be imported at levels of about 100
million pounds per year. A domestic source of hydroxy fatty
acids, say from lesquerella, would replace allergenic castor
as a source of higher nylons and other products.

The majority of the 55 Lesquerella species are native
chiefly to the arid parts of western North America (Table 2,
Chapter 2) from east central Mexico to Alberta and

Saskatchewan, in Canada (29). Some are found in the Southeast U.S., including L. globosa, L. densipila and L.
gracilis. Of particular interest for the Southeast might be
Lesquerella globosa, indigenous to central Kentucky and
Tennessee (16). In the late 1960s, genetic and agronomic
studies were carried out on experimental test plots. Seed
yields of over 1800 pounds per acre were obtained in some
cases. The seeds contained 20 to 40 percent oil having
from 50 to 74 percent hydroxy fatty acids. The hydroxy
fatty acid contents of the lesquerella seed oils are lower
than in castor oil which normally has 80 to 87 percent
hydroxy fatty acids. The seed meal obtained from the
Lesquerellas, however, resembles the meal of rapeseed and
crambe in toxicological properties and should be suitable
as an animal feed. By contrast, castor seed meal has toxi-
cants that are much more dangerous and prevent its ready
use as an animal feed (17).

Lesquerolic acid (20 carbons) resembles the 18-carbon
ricinoleic acid because both acids have only one methylene
group (CH_2) separating the double bond and the hydroxy
group. Since lesquerolic and ricinoleic acids are chemically very similar it is anticipated that chemical derivatives and products with analogous properties should be
possible. A major use for castor oil and hence for ricinoleic acid is in the production of nylon-11. By a process
analogous to nylon-11 production, it should be possible to
make nylon-13 from lesquerolic acid as indicated in Figure
4.

Another nylon that might be worthy of consideration as
a vegetable oil product is nylon-12, presently a commercial
nylon, produced petrochemically. The reaction process that
is suggested in Figure 5 is based on known chemical reactions of fatty acids. This reaction sequence has never
been carried out using lesquerolic acid, to the best of our
knowledge.

Lesquerolic acid could also be the source of undecanedioic acid, an 11-carbon diacid, and dodecanedioic acid, a
12-carbon diacid, as shown in Figure 6. There is no commercial source of undecanedioic acid. It could be used in
nylons and polyesters. Dodecanedioic acid is produced
commercially by DuPont from the trimerization of petrochemical butadiene. In 1976 it was estimated that DuPont's production capacity for dodecanedioic acid was about 40 million
pounds per year. Dodecanedioic acid is used in the manufacture of nylon-6/12 and "Qiana," and could be used for
polyesters, polyureas and other polymers (30-32). The
reaction sequences proposed in Figure 6 have never been

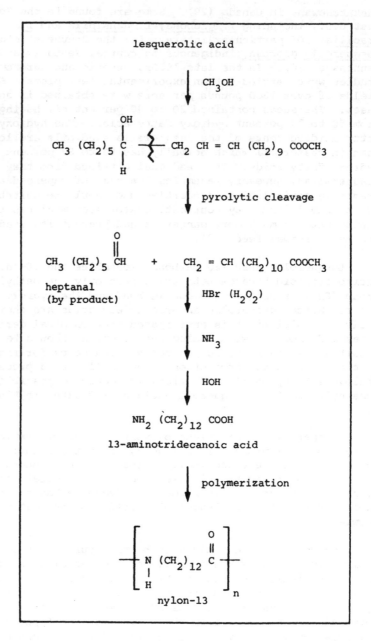

Figure 4.   Nylon-13 from lesquerolic acid (proposed)

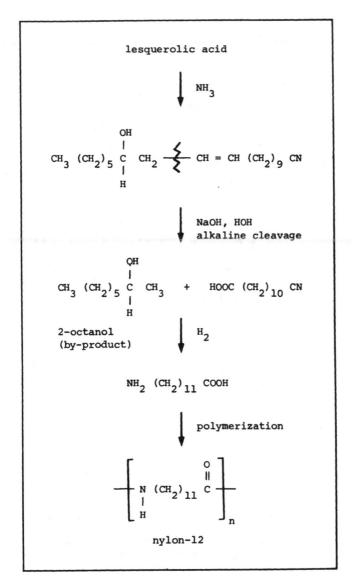

Figure 5. Nylon-12 from lesquerolic acid (proposed)

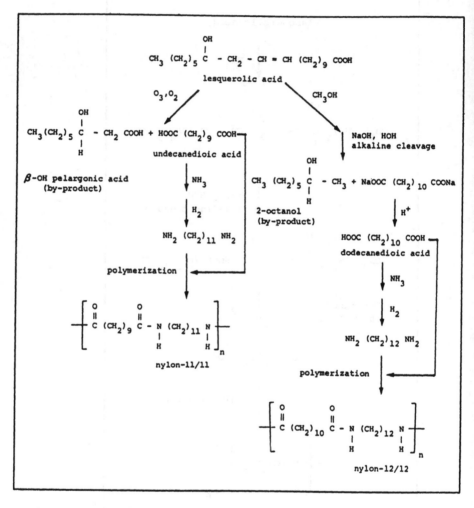

Figure 6.   Undecanedioic acid, dodecanedioic acid,
nylon-11/11 and nylon-12/12 from lesquerolic acid (proposed)

tested on lesquerolic acid, to the best of our knowledge, although they are known to work with ricinoleic acid.

The formation of other higher nylons based on 11-, 12-, and 13-carbon chain combinations could also be considered, if both erucic and lesquerolic acids were available. In short, these two acids open the possibility of production of several higher nylons that might be competitive with nylon-11 from castor oil, and petrochemical nylon-12, the only higher nylons commercially available.

Conclusions

Higher nylons can be made from erucic acid and lesquerolic acid singly, or in combination. These acids are available from novel oilseeds that are indigenous or adapted to the Southeast U.S. and have potential for development as crops. Potential erucic acid sources are Crambe abyssinica (crambe), Eruca sativa (rocket salad) and Iberis amara (rocket candytuft). Crambe, well-developed agronomically, might be applicable only to Kentucky and other northern-tier states of the Southeast. Rocket salad and rocket candytuft may be applicable in all or most southeastern states but have received less attention. Possible lesquerolic acid sources in the Southeast are Lesquerella globosa and Lesquerella densipila. Agronomic studies should be started on both, as well as Eruca sativa and Iberis amara.

Higher nylons are specialty polymers with engineering properties that tend to lead to high unit prices, in comparison with the commodity nylons based on shorter (6-carbon) chains. Only two of the many possible higher nylons are presently commercial, castor oil-derived nylon-11 and petrochemical nylon-12. The markets are well-developed. Castor oil production involves health hazards, and is not practical for the United States. This suggests the possibility of developing the aforementioned southeastern plants as sources of erucic and lesquerolic acid to make several higher nylons, including nylon-13 and 1313, as well as others, with properties similar to those of the commercial and successful nylons-11 and -12. Other nylons such as 6/12, and polyesters and polyureas may also be feasible, and the by-products may be utilized in growing markets in synthetic lubricant and plasticizer manufacture.

The Northern Regional Research Center of the U.S. Department of Agriculture has demonstrated the feasibility of producing the nylon-13s from erucic acid, and cost estimates appear favorable. Similar laboratory, pilot-plant

and economic studies need to be carried out on nylons from
lesquerolic acid.

Acknowledgements

Many members of the staff of the USDA Northern Regional
Research Center have been helpful in providing information
as background for this study. We especially thank John
Rothfus, Robert Kleiman, L. H. Princen and Everett Pryde
for helpful discussions.

References and Notes

1. Coal or oil shale as alternatives to petroleum would
   involve conversion to relatively small molecules for
   petrochemical feedstocks, such as the light aromatics
   and low-molecular-weight paraffins and olefins presently used for production of commodity petrochemicals.
   Another example of an alternative involving small
   molecules is alcohol from sugar or starch crops.

2. Fats are relatively saturated and exist as solids or
   semi-solids at room temperature; oils are more
   unsaturated (contain more double and triple bonds) and
   exist as liquids. Both are largely triglycerides.

3. Typical of the many publications emanating from the
   Northern Regional Research Center over many years are
   the following:

   a. F. R. Earle and Q. Jones, "Analyses of Seed
       Samples from 113 Plant Families," Economic Botany
       (October-December, 1962) pp. 221-250.

   b. Q. Jones and F. R. Earle, "Chemical Analyses of
       Seeds II: Oil and Protein Content of 759 Species,"
       Economic Botany (April-June, 1966) pp. 127-155.

   c. A. S. Barclay and F. R. Earle, "Chemical Analyses
       of Seeds III: Oil and Protein Content of 1253
       Species," Economic Botany (April-June, 1974)
       pp. 179-236.

   d. C. R. Smith, Jr., "Unusual Seed Oils and Their
       Fatty Acids," Fatty Acids, ed. E. H. Pryde,
       American Oil Chemist's Society, Champaign, IL,
       (1979) pp. 29-47.

e. L. H. Princen, "New Crop Development for Industrial Oils," <u>Journal of American Oil Chemists' Society</u> (September, 1979), pp. 845-848.

f. R. Kleiman and G. F. Spencer, "Search for New Industrial Oils: XVI. Umbelliflorae - Seed Oils Rich in Petroselenic Acid." <u>Journal American Oil Chemists' Society 59</u> (1982) pp. 29-38.

4. R. P. Morgan and E. B. Shultz, Jr., "Fuels and Chemicals from Novel Seed Oils," <u>Chemical and Engineering News</u> (September 7, 1981) pp. 69-77.

5. I. A. Wolff, "Seed Lipids," <u>Science</u> 154 (December 1966) pp. 1140-1149.

6. E. W. Eckey, <u>Vegetable Fats and Oils</u>, Reinhold Publishing Co., New York, N.Y. (1954), pp. 19-70.

7. "Statistics: Yearly Summary, 1980," <u>Fatty Acid Monthly Reports</u>, Fatty Acids Producers' Council, 475 Park Avenue South at 32nd Street, New York, N.Y., 10016.

8. A. G. Johanson, "Historical and Marketing Trends of Natural/Synthetic Fatty Acids," <u>Journal of the American Oil Chemists' Society</u> (November, 1977) pp. 848A-825A.

9. E. H. Pryde, "Fats and Oils as Chemical Intermediates: Present and Future Uses," <u>Journal of American Oil Chemists' Society</u> (September, 1979), pp. 849-854.

10. L. W. Codd, ed., <u>Chemical Technology: An Encyclopedic Treatment</u>, Vol. 6, Barnes & Noble Books (New York, 1973) pp. 570-577.

11. "Materials 1981: The Statistics," <u>Modern Plastics</u> (January, 1981) pg. 67.

12. a. <u>Rilsan: Design Guide to a Versatile Engineering Plastic</u>, Technical brochure from Rilsan Industrial Inc., 1112 Lincoln Rd., Birdsboro, PA., 19508.

    b. <u>Rilsan</u>, Brochure from Rilsan Industrial Inc., 1112 Lincoln Rd., Birdsboro, PA., 19508.

13. M. I. Kohan, ed., <u>Nylon Plastics</u>, John Wiley & Sons, New York, N.Y. (1973) pp. 20-29.

14. J. Kestler, "What About the 'Other Nylons?'," <u>Modern Plastics</u> (August, 1968) pp. 86-89.

15. a. R. J. Youle and A. H. C. Huang, "Evidence That the
         Castor Bean Allergens are the Albumin Storage
         Proteins in the Protein Bodies of Castor Bean,"
         Plant Phys. 61 (1978) pp. 1040-1042.

     b. E. M. Apen, Jr., W. C. Cooper, R. J. M. Horton and
         L. D. Scheel, Health Aspects of Castor Bean Dust,
         U.S. Dept. of Health, Education and Welfare,
         Public Health Service (1967).

16. H. M. Draper, III, and E. B. Shultz, Jr., "New Oil-
     seed Crops for Specialty Chemicals in the Southeastern
     United States," Presented at the Annual Meeting of the
     Society for Economic Botany, University, Alabama,
     June 14-16, 1982.

17. L. H. Princen, "New Crop Developments for Industrial
     Oils," Journal of American Oil Chemists' Society
     (September, 1979) pp. 845-848.

18. E. H. Pryde, "Vegetable Oil Raw Materials," Journal of
     American Oil Chemists' Society (November, 1979) pp.
     719A-725A.

19. H. J. Nieschlag and I. A. Wolff, "Industrial Uses of
     High Erucic Oils," Journal of American Oil Chemists'
     Society (November, 1971) pp. 723-727.

20. R. K. Downey, "Agricultural and Genetic Potential of
     Cruciferous Oilseed Crops," Journal of American Oil
     Chemists' Society (November, 1971) pp. 718-722.

21. F. R. Earle, J. E. Peters, I. A. Wolff and G. A. White,
     "Compositional Differences Among Crambe Samples and
     Between Seed Components," Journal of American Oil
     Chemists' Society (May, 1966) pp. 330-333.

22. K. J. Lessman and W. P. Anderson, "Crambe," in New
     Sources of Fats and Oils, E. H. Pryde, L. H. Princen
     and K. D. Mukherjee eds., American Oil Chemists'
     Society, Champaign, IL., (1981) pp. 171-176.

23. J. L. Greene, R. E. Burks, Jr. and I. A. Wolff,
     "13-Aminotridecanoic Acid from Erucic Acid," Industrial
     and Engineering Chemical Prod. Research and Development
     (June, 1969) pp. 171-176.

24. K. D. Carlson, V. E. Sohns, R. B. Perkins, Jr. and
     E. L. Huffman, "Brassylic Acid: Chemical Intermediate
     from High Erucic Oils," Industrial and Engineering

Chemical Product Research and Development (January, 1977) pp. 95-101.

25. H. J. Nieschlag, J. A. Rothfus, V. E. Sohns and R. B. Perkins, Jr., "Nylon-1313 from Brassylic Acid," Industrial and Engineering Chemical Product Research and Development (January, 1977) pp. 101-107.

26. R. B. Perkins, J. J. Roden, III, A. C. Tanquary, and I. A. Wolff, "Nylons from Vegetable Oils: -13, -13/13 and -6/13," Modern Plastics (May, 1969) pp. 136-142.

27. V. E. Sohns, "Cost Analyses for New Products and Processes Developed in USDA Laboratories," Journal of American Oil Chemists' Society (September, 1971) pp. 363A-384A.

28. A-M. V. Watowich, Market Potential for Commercialization of Polymers, Plasticizers and Lubricants from Unusual Oilseeds, M.S. Thesis, Dept. of Technology and Human Affairs, School of Engineering, Washington University, St. Louis, MO (May, 1982).

29. R. W. Miller, C. H. Van Etten and I. A. Wolff, "Amino Acid Composition of Lesquerella Seed Meals," Journal of American Oil Chemists' Society (February, 1962) pp. 115-117.

30. E. C. Leonard, "The Higher Aliphatic Di- and Polycarboxylic Acids," in Fatty Acids, E. H. Pryde, ed., American Oil Chemists' Society, Champaign, IL, (1979) pp. 504-526.

31. R. G. Kadesh, "Fat-Based Dibasic Acids," Journal of American Oil Chemists' Society (November, 1979) pp. 845A-849A.

32. E. H. Pryde and J. C. Cowan, "The Aliphatic Dibasic Acids," in High Polymers Vol. 27: Condensation Monomers, J. K. Stille and T. W. Campbell, eds., Wiley-Interscience, New York (1972) p. 82.

8. Sunflower Oil as an Alternative Farm Fuel

Introduction

Our view of energy is very different today than it was 10 years ago. During the 1970's, Americans had to face some harsh truths about their energy future. The beginning of the decade saw declining production of domestic oil and natural gas combined with increasing oil imports. Today, almost 45% of the total energy of the United States is derived from petroleum oil, 40% of which is imported (1) and the majority of this imported oil comes from an internationally traded oil pool controlled by The Organization of Petroleum Exporting Countries (OPEC). An assured economical foreign oil supply is a thing of the past, and it is now abundantly clear that the United States must solve its energy problem at home. Various alternate liquid fuels are being considered, including vegetable oils.

Vegetable Oil Production

Cost Comparisons

One way to compare costs of fuels is to look at the heating values on a monetary basis. Petroleum-based fuels are still the most economic as shown in Table 1. Alternate fuels are not yet economically competitive with petroleum-based fuels. However, when one considers only the alternative fuels, sunflower fuel is more attractive than ethanol.

Even though vegetable oils are not yet economically competitive, they may be so in the future. The price ratios of two vegetable oils and No. 2 diesel fuel for the past 10 years are presented in Figure 1. It appears that the price differential between vegetable oils and diesel fuel is decreasing. On the other hand, it can also be seen in Figure 1 that this price ratio is sensitive to political events such as the 1973-74 oil embargo.

Table 1. Fuel Cost Comparisons, November 1982

	Cost per liter, dollars	High Heating Value kJ/liter	Cost per GJ, dollars
Unleaded Gasoline	0.35	34,800	10.06
No. 2 Diesel Fuel	0.31	38,400	8.07
Ethanol (200 proof)	0.45	23,500	19.13
Sunflower Oil (crude)	0.41	36,600	11.20

Sources: DOE. Fuel From Farms. Technical Information Center, U.S. DOE, Oak Ridge, Tennessee. February 1980.

K.R. Kaufman, M. Ziejewski, M. Marohl, H. Kucera and A.E. Jones. Performance of diesel oil and sunflower oil mixtures in diesel farm tractors. Paper #81-1054. American Society of Agricultural Engineers, St. Joseph, MI. 1981.

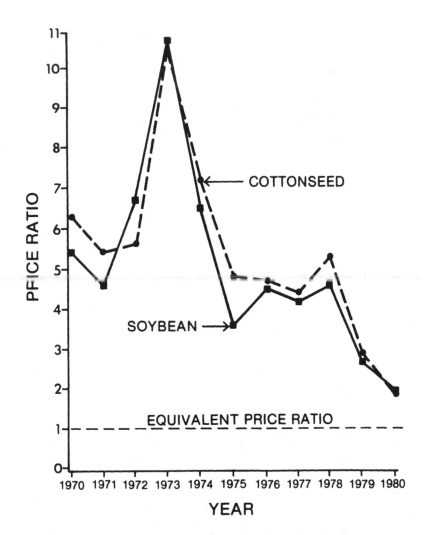

Fig. 1. Price ratio of vegetable oil to on-farm diesel fuel.

Sources: USDA. 1975. Agricultural prices annual summary 1974. Crop Reporting Board, Statistical Reporting Service; USDA, Washington, D.C. June.

USDA. 1977-1982. Fats and oils situation. FOS-289 thru FOS 307. Economic Research Service, USDA, Washington, D.C.

USDA. 1981. Agricultural prices annual summary 1980. Crop Reporting Board; USDA, Washington, D.C. June.

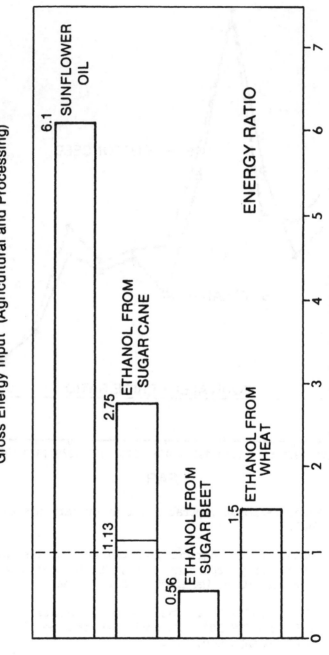

RATIO OF $\dfrac{\text{Total Energy in Fuel}}{\text{Gross Energy Input (Agricultural and Processing)}}$

SUNFLOWER OIL   6.1

ETHANOL FROM SUGAR CANE   2.75

ETHANOL FROM SUGAR BEET   1.13

ENERGY RATIO

ETHANOL FROM WHEAT   1.5

0.56

0   1   2   3   4   5   6   7

Fig. 2. Energy Ratio. Reprinted by permission from G.R. Quick, "Farm Fuel Alternatives,"
Power Farming (Australia) 10-17 Feb. 1980.

Energy Balance

The energy balance is another term of interest when alternate fuels are discussed. A comparison was made by Quick (2) and is shown in Figure 2. Vegetable oils are an excellent producer of energy. The overall efficiency for sunflower oil is the highest of all the farm fuel alternatives. This efficiency is based on the ratio of fuel output to agriculture plus processing energy inputs.

A similar energy analysis was performed by Schoedder (3) for the agricultural production and processing of 1 hectare of rapeseed oil. He stated that it requires 21.95 GJ/ha energy equivalent of inputs for the agricultural production of rapeseed. The crop yields 64.7 GJ/ha of rapeseed and 59.8 GJ/ha of rape straw for a total energy content of 124.5 GJ/ha. Nearly 50% of the crop energy (59.8 of the 124.5 GJ/ha) is plowed down in the field. An energy flow diagram is shown in Figure 3.

For the separation of rapeseed (64.7 GJ/ha) into rape oil (39.5 GJ/ha) and rape cake (25.2 GJ/ha), energy equivalents of 0.65 GJ/ha for the oil mill plant and 3.4 GJ/ha for process energy are needed (3). It follows from this that for the production of raw rape oil and rape cake there is an energetic output/input relation of 2.49. An additional input of 1.15 GJ/ha is required for refining the raw rape oil. So the total production of purified rape oil is characterized by an energy output/input ratio of 2.24. If the high value energy invested by technical means (22.6 GJ/ha) is subtracted from the energy of the purified oil (37.5 GJ/ha), then a net energy of 14.9 GJ/ha is yielded, e.g. the equivalent of 411 l/ha of diesel fuel. It is assumed in this calculation that the process energy (5.2 GJ/ha) can be taken from another source of lower value energy; on the other side, the energy of rape cake (25.2 GJ/ha) has not been considered. In comparison to other processes for fuel from renewable agricultural resources, rape oil production gives good output/input ratio and considerable net output of fuel energy but does not give high amounts of energy per unit of area.

Agronomic Aspects

The conventional vegetable oils that could be considered for use as alternate fuels include sunflower, soybean, peanut, cottonseed, rapeseed, safflower, and corn. Data on most of these crops are presented in Table 2. Information on rapeseed and safflower was not available since they are small and localized crops in the United States. Of the other crops, soybeans

Table 2. Area, Yield and Production of Selected Products, United States, 1978-1980

	Sunflowers	Soybeans	Peanuts	Cottonseed	Corn
Area Harvested (Million Hectares)					
1978	1.13	25.76	0.61	5.02	29.11
1979	2.19	28.56	0.62	5.19	29.30
1980	1.52	27.46	0.57	5.26	29.57
Crop Yield (Metric Tons Per Hectare)					
1978	1.53	1.96	2.94	0.77	6.34
1979	1.51	2.16	2.93	1.01	6.88
1980	1.14	1.80	1.83	0.75	5.71
Total Production (Million Tons)					
1978	1.73	50.86	1.79	3.87	184.61
1979	3.31	61.72	1.80	5.24	201.66
1980	1.73	49.45	1.04	3.96	168.85
Vegetable Oil Production (Liters Per Hectare)					
1978	663	384	1010	134	344
1979	655	421	1007	175	373
1980	494	351	629	130	309
Production Costs, Excluding Land (Dollars Per Hectare)					
1978	---	244.95	933.83	647.71	371.22
1979	222.57	284.15	1033.63	784.70	439.48
1980	263.48	324.87	1137.27	862.47	527.67

Sources: USDA. Farmline, Vol. II, No. 7, USDA Economics and Statistics
Service. Washington, D.C. August 1981.

USDA. Crop production. 1980 Annual Summary. USDA ESSS
Crop Reporting Board. Washington, D.C. January 14, 1981.

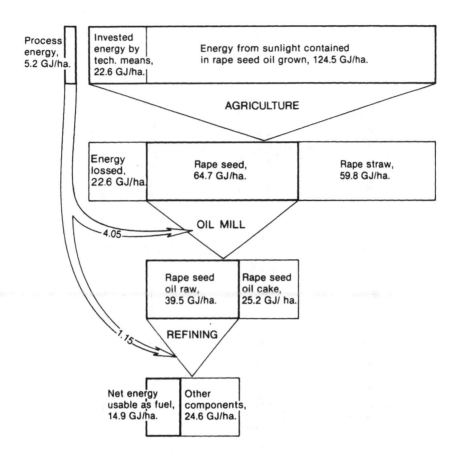

Fig. 3. Energy flow diagram for production of purified rapeseed oil. Reprinted by permission from <u>Beyond the Energy Crisis</u>, Vol. III, Fazzolare and Smith, eds., Pergamon Press, 1981.

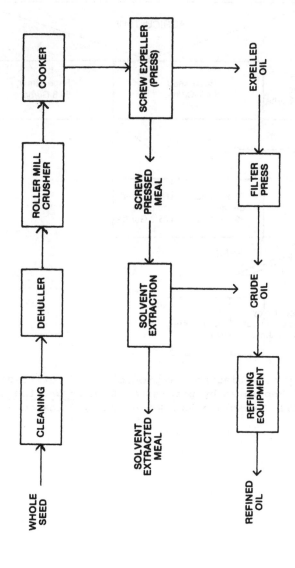

Figure 4. Commercial Sunflower Oil Extraction

and corn are produced in the largest quantity and are generally available over the largest geographical area. Peanuts and
sunflowers produce the largest yields of oil per hectare, but
the production costs of sunflowers are significantly lower
than those for peanuts. Therefore, from an agronomic view-
point, sunflowers seem to be the most promising vegetable oil
candidate for an alternate fuel.

The practicality of an alternate farm fuel is dependent
on the amount of land required to produce the crop. The average sunflower yield in the United States from 1978 through
1980 was 1.39 metric tons per hectare (1243 pounds per acre).
The oil content of the sunflower seed is 40%. Almost all the
oil in the sunflower seed can be extracted in a commercial operation. It is possible to produce 604 liters of sunflower oil
per hectare (65 gallons per acre) of land. The on-farm fuel
required to produce 1 hectare of sunflowers or small grain in
North Dakota ranges from 56 to 84 liters per hectare (6 to 9
gallons per acre) (4). Under these conditions, 1 hectare of
sunflowers could produce enough fuel to grow 7 to 11 hectares
of small grain or sunflowers. About one-tenth of a farmer's
land devoted to the production of sunflowers could provide his
on-farm fuel requirements. In 1919, the agricultural acreage
devoted to the production of feed for horses and mules in the
United States was 22 percent of the harvested cropland (5).
We should not be surprised if once again a sizeable percentage
of our farmland will be devoted to the production of raw
materials for the manufacture of fuel.

Processing of Sunflower Seed

Commercial Extraction

Technology for commercial extraction of oil from seeds is
well developed. A simplified flow diagram of a typical commercial process is shown in Figure 4. The seeds need to be
stored under proper conditions to maintain the quality of the
seed. The seed is first cleaned; stones or metal pieces will
damage the expeller. The hulls are removed to reduce the fiber content and increase the protein content of the meal.
The whole seed is then crushed and heated to obtain maximum
oil recovery. The press separates the oil from the meal.

Most existing oil extraction plants include auger-type
expeller units. The machines are provided with an auger in a
perforated housing. The center shaft of the auger increases
in size from inlet to outlet. The outlet end of the shaft is
the same diameter as the flighting. As seed moves through the
auger, it is compressed between the shaft and the auger hous-

ing. Oil is forced out of the seed and flows from the machine
through the perforations in the housing as it is compressed.
The amount of oil remaining in the seed varies. The expelled
oil is filtered to remove small seed particles and meal fines.

The meal from the screw press is exposed to a spray of
hexane, a solvent which dissolves the remainder of the oil in
the meal and removes it from the meal to increase the oil
yield. Over 99% of the oil in the seed can be removed when
this process is used. Hexane is evaporated from the oil and
reused. The resulting crude oil contains the triglyceride
sunflower oil plus free fatty acids and other fat-like material referred to as "gums". Further refining is done to
obtain oil suitable for human consumption.

Feeding Trials of Sunflower Meal (Solvent Extracted).
Three types of sunflower meals are generally available: a 42%
protein meal from dehulled seed; a 34% to 35% protein meal
from partially dehulled seed; and 28% protein meal from seeds
with hulls. Each of these meals are both press and solvent
extracted and usually have less than 2% residual oil. The
sunflower meal protein content compares favorably with soybean, cottonseed, and rapeseed meals.

Meal from processed sunflowers shows promise for being
a useful feed supplement for livestock (6). Dr. W.E. Dinusson
is conducting a research program in the Department of Animal
Science at North Dakota State University to evaluate this
meal. In experiments with cattle and sheep, the sunflower
meals have given results similar to soybean meal when substituted on an equal protein basis. The energy values of the
sunflower meals are less than that of soybean meal because of
added fiber content when some or all of the hulls are included.

In swine rations, the limiting factor in sunflower meal
is the lack of the amino acid lysine (6). In the high protein meals, 42% or more crude protein from dehulled seeds, the
lysine content is only 1.5% compared to about 3% in soybean
meal. In the 24% protein meal, the lysine content is about
1.2% and in the 28% meal it is about 1.0%. Substituting sunflower meal for 1/3 or 2/3 of the soybean meal and adding
synthetic lysine to equal the lysine provided by the soybean
meal gave very similar results to soybean meal in barley-based
rations for swine growing-finishing rations. Substituting all
the soybean meal with sunflower meal, even with the lysine
additions, gave more variable results. However, the differences were small.

In corn-based rations for swine, 1/3, 2/3 or all the soybean meal could be replaced if lysine was added to make up

the difference between the sunflower meal and soybean meal rations. Lysine supplement is expensive, so when all the soybean meal was replaced in corn-based rations the cost of the lysine was equal to or greater than the cost of the sunflower meal supplied to furnish the necessary protein.

Fertilizer Value of Sunflower Meal (Solvent Extracted). The meal has also been evaluated for possible use as a fertilizer by Dr. Edward J. Diebert in the Department of Soils at North Dakota State University (7). Fertilizer prices of 40, 53, and 26 cents per kilogram (18, 24, and 12 cents per pound) respectively for nitrogen, phosphorus, and potassium were used to calculate the value of a metric ton of sunflower meal. Both whole seed meal and dehulled meal were considered. These are typical solvent extraction products. The fertilizer content of the whole seed meal and dehulled meal was calculated to have an approximate value of $32 and $57 per metric ton ($29 and $25 per ton) respectively. The meal could be effectively utilized as a supplemental fertilizer source when other means of disposal are not available.

Small-Scale Extraction

Small-scale production of sunflower seed oil is conceivable. Figure 5 shows the steps involved in small-scale processing of sunflower seed. Several of the steps involved in a commercial operation have been eliminated including the solvent extraction process which uses hexane, a highly flammable solvent. The safety problems associated with solvent extraction could be serious.

The by-product meal from a small-scale extraction process differs from the solvent-extracted meal. The important difference is that the screw-pressed meal contains residual oil. The amount of oil varies from 7% to 20% on a dry weight basis compared to less than 2% for solvent-extracted meal. Table 3 shows the oil yield in liters per hectare from a small-scale extraction process.

Table 3. Oil Yield in l/ha From a Small-Scale Process

% of Oil Extracted	Seed Yield (t/ha)			
	1.0	1.4	1.8	2.2
60	260	364	468	572
80	347	486	624	573

Assumed sunflower seed oil content = 40% by weight.

The meal also has a high fiber content since the hulls have not been removed.

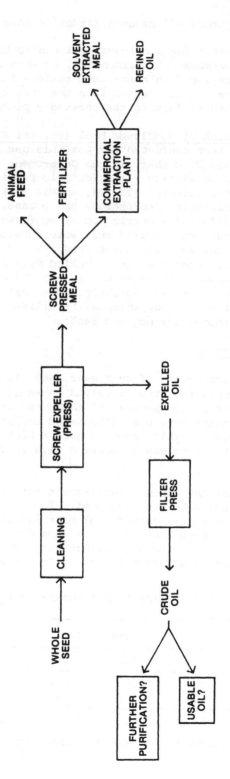

Figure 5. Small-Scale Sunflower Oil Extraction

Feeding Trials of Sunflower Meal(Screw Pressed). Feeding in excess of 0.5 kilogram (1 pound) of oil daily (2.5
kilograms daily of the screw press meal) to 196 kilogram (433
pound) dairy steers has resulted in loss of weight and scours
(8). The meal, when fed in excess, also seems to reduce digestibility of protein and fiber in the rations. This meal is
considerably higher in fat content than meal from a solvent
extraction process and should be used quickly to avoid
rancidity.

Fertilizer Value of Sunflower Meal (Screw Pressed). The
meal could be used as a fertilizer (9). Analysis showed 3.66%
nitrogen (N), 1.10% phosphorus (P) and 1.62% potassium (K) or
approximately 37, 11 and 16 kilograms of nutrient respectively
per metric ton of meal. Based on current fertilizer prices
(N @ 51 cents/kg, P @ 115 cents/kg and K @ 29 cents/kg) and
sunflower meal value ($121.00/t for 28% protein), the meal has
more value as a protein feed supplement than as a fertilizer
source ($35.81/t). This does not take into account the bene-
ficial effect of the meal on the physical condition of the
soil. Also, the sunflower meal would have some residual
nutrient benefit once the undecomposed meal is mineralized in
the soil. A three-fold increase in current fertilizer prices
would make the use of sunflower meal as a soil additive more
competitive. The use of sunflower meal as a fertilizer source
appears to be an alternative method for utilization of this
by-product material as evidenced by the response on two soils
tested. The response obtained was greatest on the soil with
lower N, P, K, and organic matter soil test levels suggesting
that maximum benefit would be obtained by applying the sunflower meal to soils of this fertility status.

Refining of Vegetable Oils. The oil extracted from the
clean seed contains small seed particles and meal fines which
settle to the bottom of a holding tank after a few days. The
sunflower oil is skimmed off the top of this tank and run
through a filter. The filtration is necessary to prevent
plugging of fuel filters on the diesel engine. The filter
should remove all particles greater than 5 micrometers in
diameter.

After settling and filtering, the oil contains triglycerides, free-fatty acids and other fat-like material. The
latter may cause gums to form in engines,so further processing
is required before the oil is satisfactory for diesel engines.
The levels of refining of vegetable oils may be defined as
follows (10):

1.   Crude oil: raw oil removed from the vegetable oil
      seed and filtered. The extraction method used

should be indicated; i.e., screw press, solvent extraction, etc.

2. Crude degummed oil: crude oil which has most of its gum removed. A small quantity of water is intimately mixed to hydrate the gums, which are then removed by centrifuging, settling, or cooling. This is the minimum recommended level of refining for diesel engine fuel.

3. Alkali refined oil: oil which has been chemically treated, usually alkali refined and mechanically or physically separated using centrifugation, to remove the free-fatty acids and gums.

4. Fully refined oil: produced from alkali refined oil by bleaching, winterizing (cooling of oil to produce partial crystallization followed by separation of shortening or margarine solids for high quality salad oils), and deodorizing. This oil is suitable for human consumption. Bleaching and deodorizing are not needed for diesel engine fuel.

Economic Analysis. Costs were estimated for processing a volume of sunflower oil sufficient to provide the fuel requirements for an average size farm in North Dakota (11). An average North Dakota farm is estimated to use approximately 18,200 liters (4800 gallons) of diesel fuel annually, assuming diesel engines were used for all field, harvest and livestock operations. Three sizes of screw presses were analyzed for on-farm usage: 0.32 metric tons of whole sunflower seed per day (0.35 tons/day), 1.51 metric tons per day (1.67 tons/day), and 4.54 metric tons per day (5.0 tons/day). The assumption was used that presses would operate eight hours per day plus an additional hour for starting and shutting the press down. The small press had an oil extraction efficiency of 82 percent while the medium and large presses had extraction efficiencies of 89 percent. These presses extract sunflower oil from whole sunflower seed without any cooking or heating process which makes the oil extraction efficiency lower than that from commercial presses, resulting in meal with a higher oil content. Total estimated costs of crude sunflower oil came to $1.14 per liter ($4.33/gallon) for the 0.32 metric ton/day press, $0.76 per liter ($2.82 gallon) for the 1.51 t/day press, and $0.75 per liter ($2.87/gallon) for the 4.54 t/day press. These costs include a charge for all the resources used in producing the crude sunflower oil. Total time required to produce 18,200 liters (4800 gallons) of crude sunflower oil varied from 1440 hours for the smallest press , 280 hours for the medium sized press,to only 93 hours for the largest press. Translated into nine-hour operating days, this represents 160 days or 5.3 months for the smallest press, 31 days for the medium sized press and 10 days or about one-third of a month for the largest press.

As a means of providing for a wide range of alternatives in analyzing the cost structure, it was assumed that some farmers may already have a building suitable to house the press and a zero opportunity cost was considered for the operator's labor. By deleting building depreciation and the labor charge, the resulting cost per liter of crude sunflower oil is $0.68 ($2.57/gallon) for the smallest press, $0.60 ($2.27/gallon) for the medium press, and $0.66 ($2.48/gallon) for the large press.

When the on-farm sized presses are operated 300 nine-hour days per year, (eight hours for the press and one hour to start and stop the press at the beginning and end of the day), the volume of crude oil processed totals 34,068 liters (9,000 gallons) for the small press, 175,160 liters (46,272 gallons) for the medium press, and 525,022 liters (138,696 gallons) for the largest press. The processing costs per liter decrease when the presses can be operated full time. The total estimated costs of the crude sunflower oil from the three presses operating 300 days per year was $1.04 per liter ($3.93/gallon) for the 0.32 t/day press, $0.54 per liter ($2.03/gallon) for the 1.51 t/day press, and $0.46 per liter ($1.74/gallon) for the 4.54 t/day press. These costs include a charge for all resources. By deleting building depreciation and the labor charge, the costs are $0.79 per liter ($2.99/ gallon) for the small press, $0.52 per liter ($1.97/gallon) for the medium press, and $0.46 per liter ($1.73/gallon) for the large press.

Conclusions

The American dependence on fossil fuels has dictated that the United States start looking for alternate fuels. Various alternate liquid fuels are being considered,including vegetable oils. One of the benefits of growing plants for oil is the wide variety of species available. One can choose the plant that is best suited for the growing conditions in a particular area. About 10 percent of a farmer's land devoted to the production of sunflower could provide his direct on-farm fuel requirements.

Before we turn to vegetable oils as fuels, we need to know how much they cost and how much energy it takes to grow and process these plants. Vegetable oil production gives a good energy input/output ratio with considerable net fuel energy output. Vegetable oil fuels are not economically competitive with diesel fuel but the price differential is decreasing. On-farm processing is more expensive than commercial production of vegetable oil fuels.

Vegetable oils have an advantage in that it is easy to extract the oil from them by simple mechanical operations. They can be processed on an industrial scale by a pre-press solvent extraction process. The meal by-product can be utilized as a high-protein animal feed or as a fertilizer. Seed can also be processed on a small scale by a screw press; however, this process is not as efficient and the meal by-product contains residual oil. This meal can be used as an animal feed, a fertilizer, or can be further processed commercially. The by-product meal from the small-scale process is more valuable as an animal feed than as a fertilizer but it should be fed in limited quantities.

References

1. Anonymous. The national energy outlook 1980-1990. Shell Oil Company, Houston, TX. August, 1980.

2. Quick, G.R. 1980. "Farm Fuel Alternatives", Power Farming Feb. 10-17, 1980.

3. F. Schoedder, "Rape Seed Oil as an Alternative Fuel for Agriculture". In Beyond the Energy Crisis. Vol. III. Fazzolare and Smith, Eds. (Pergamon Press, Oxford, England. 1981) pp. 1815-1822.

4. R.R. Frith and W.J. Promersberger. "Estimates of Fuel Consumption for Farming and Ranching Operations Under Typical North Dakota Conditions", Bulletin 493. Agricultural Experiment Station, North Dakota State University, Fargo, ND 1973.

5. D. Bartholomew. "Vegetable Oil Fuel," Journal of the American Oil Chemists Society, Vol. 58, No. 4 (1981) pp. 286A-288A.

6. G.L. Pratt, L. Backer, K. Kaufman, L. Jacobsen, C. Olson, P. Ramdeen, W. Dinusson, D. Helgeson, L. Schaffner, and H. Klosterman. "On-Farm Production of Sunflower Oil for Fuel." In Beyond the Energy Crisis, Vol. III. Fazzolare and Smith, Eds. (Pergamon Press, Oxford, England, 1981) pp. 1767-1774.

7. V.L. Hofman, W. Dinusson, D. Zimmerman, D. Helgeson and C. Fanning. "Sunflower Oil as a Fuel Alternative." Circular AE 694. Cooperative Extension Service, North Dakota State University, Fargo, ND, 1980.

8. W.E. Dinusson, L.J. Johnson, and R.B. Danielson. "Squeezing" - Sun Oil Meal for Cattle. North Dakota Farm Research. Vol. 39, No. 6, North Dakota State University, Fargo, ND, May-June, 1982, pp. 25-26.

9. E.J. Diebert, and D. Lizotte. "Soil Applications of Sunflower Meal as Potential Fertilizer Sources. North Dakota Farm Research. Vol. 39, No. 6, North Dakota State University, Fargo, ND, May-June, 1982, pp. 15-17.

10. D.R. Erickson. Soy oil primer (American Soybean Association, St. Louis, MO. July 9, 1979).

11. D. L. Helgeson and L.W. Schaffner. "The Economics of On-Farm Processing of Sunflower Oil". North Dakota Farm Research. Vol. 39, No. 4, North Dakota State University, Fargo, ND, Jan-Feb., 1982. pp. 3-7.

9. Testing of Vegetable Oils in Diesel Engines

Introduction

The world petroleum situation of the past several years has focused attention on the need for development of alternate fuels. It has been recognized for many years that vegetable oils can be burned in compression ignition engines. As early as 1900, a diesel engine was demonstrated running wholly on ground-nut oil at the Paris exposition (1). However, vegetable oils have not been accepted as viable substitutes for diesel fuel because of inexpensive and abundant supplies of petroleum-based fuels.

Fuel Properties

Diesel Fuel

Almost all of the fuels commonly used in diesel engines today are products of crude petroleum. In general, all crude petroleum has a molecular structure made up mainly of combined carbon and hydrogen.

Requirements for diesel fuel oils have been issued by the American Society for Testing Materials (ASTM). The two grades of diesel fuel most commonly used in modern high speed diesel engines are No. 1-D and No. 2-D. Table 1 covers the limiting requirements for these grades.

The requirements for a good compression ignition (CI) fuel cannot be simply stated. Obert states (2) "This situation arises because of the added complexity of the CI engine from its heterogenous combustion process, which is strongly affected by injection characteristics." However, he gives the following general observations which can assist in identifying good CI fuels:

Table 1. Limiting Requirements for Diesel Fuel Oils.

Grade of Diesel Fuel Oil	Flash Point C	Pour Point C	Distillation Temperatures, C, 90% Point		Viscosity at 37.8°C Kinematic,CS (SUS)		Cetane No.[c]
	Min.	Max.	Min.	Max.	Min.	Max.	Min.
No. 1-D: A volatile distillate fuel oil for engines in service requiring frequent speed and load changes.	37.8 or	a	–	287.8	14	2.5 (34.4)	40
No. 2-D: A distillate fuel oil of lower volatility for engines in indus- trial and heavy mobil service	51.7 or	a	282.2[b]	338	2.0[b] (32.y)	4.3 (40.1)	40

Source: Adapted from ASTM Specifications for Petroleum Products, ASTM D975-78, American Society for Testing and Materials, 1978.

[a]For cold weather operation, the pour point should be specified 5.6°C below the ambient temperature at which the engine is to be operated except where fuel oil heating facilities are provided.

[b]When a pour point of less than -17.8°C is specified, the minimum viscosity shall be 1.8 CS. (32.0s, Saybold Universal) and the minimum 90% point shall be waived.

[c]Where cetane number by test method ASTM D613 is not available, ASTM D976 may be used as an approximation. Where there is disagreement, ASTM D613 shall be the referee method.

1. Knock Characteristics. "The present-day measure
    is the cetane rating - the best fuel, in general,
    will have a cetane rating sufficiently high to
    avoid objectionable knock."

2. Starting Characteristics. "The fuel should start
    the engine easily. This requirement demands high
    volatility, to form a readily combustible mixture;
    and a high cetane rating in order that the self-
    ignition temperature will be low."

3. Smoke and Odor. "The fuel should not promote
    either smoke or odor from the exhaust pipe. In
    general, good volatility is demanded as the first
    prerequisite to insure good mixing and therefore
    complete combustion."

4. Corrosion and Wear. "The fuel should not cause
    corrosion before combustion, or corrosion and wear
    after combustion. These requirements appear to be
    directly related to the sulphur, ash, and residue
    contents of the fuel."

5. Handling Ease. "The fuel should be a liquid that
    will readily flow under all conditions that will be
    encountered. This requirement is measured by the
    pour point and the viscosity of the fuel. The
    fuel should also have a high flash point since an
    advantage of the CI engine is its use of fuels with
    low fire hazards."

Fuel Properties and Their Significance

ASTM has established standardized tests for diesel fuels,
evaluating the properties and characteristics which have
effects on engine performance and reliability. Discussion
follows on these properties, characteristics, and their significance to a diesel engine.

Ignition quality expressed by cetane number, relates to
ease of ignition of the fuel. Satisfactory diesel combustion
demands self ignition of the fuel as it is sprayed into the
hot swirling, compressed cylinder gas. Fuels with better
ignition quality have shorter ignition delay periods, resulting in easier starting and less knock. The higher the cetane
number, the shorter the ignition period, and the smaller the
amount of fuel in the combustion chamber upon ignition.
Consequently, high cetane-number fuels generally cause lower
rates of pressure rise and lower peak pressures, lessening
combustion noise and permitting improved control of combustion.

This results in increased engine efficiency. Higher cetane-
number fuels lead to easier starting in cold weather, faster
warm-up, and less exhaust smoke and odor. On the other hand,
extremely short delays initiate burning quickly and may inter-
fere with proper mixing and combustion of the bulk of the fuel
spray.

Distillation temperatures are those at which portions of
a fuel sample are vaporizable, and the distillation range is
inversely related to the volatility of a fuel. More volatile
fuels distill in lower temperature ranges. The significance
of volatility has been given in a Southwest Research Institute report (3):

> "Fuels having too low volatility tend to reduce
> power output and fuel economy through poor
> atomization, while those having too high volatility
> may reduce power output and fuel economy through
> vapor lock in the fuel system or inadequate
> droplet penetration from the nozzle. In general,
> the distillation range should be as low as
> possible without adversely affecting the flash
> point, burning quality, heat content, or
> viscosity of the fuel."

Viscosity is a measure of resistance to flow. Fuel
viscosity exerts a strong influence on the shape of the fuel
spray. High viscosities lead to poor atomization, large
droplets, and solid streams of fuel rather than a spray of
small droplets. As a result, fuel is not adequately mixed
with the air. Poor combustion, and loss of power and economy
ensue. In small engines, the fuel spray may impinge on
cylinder walls, washing away the lubricating oil film, causing dilution of the crankcase oil with fuel, which contributes
to excessive wear. Low viscosities, on the other hand, may
lead to poor penetration into the combustion chamber and poor
performance. Also, low-viscosity fuels have poor lubricating
properties and this may increase wear of fuel system components.

Heat of combustion or calorific value is the heat produced when the fuel is burned completely. There are two calorific values: gross and net. The difference is the heat
attained by condensing the water vapor formed by combustion.
The gross calorific value includes this energy, and the net
calorific value does not. The heat energy per unit volume of
a fuel should be high to reduce the quantity of fuel handled
by the injection system and to maximize vehicle range.

Pour point is the lowest temperature at which fluid

movement can be detected. The cloud point is the temperature
at which paraffin wax or other solids begin to crystallize
when the fuel is chilled. The cloud point is usually several
degrees higher than the pour point. Both values are important
at low temperatures. If the ambient temperature falls below
the pour point, fuel will not flow through the fuel system.
According to Obert (2): "The oil should have a pour point
10° to 15°F below the operating temperature."

Sulfur content is related to engine wear, corrosion, and
crankcase contamination. Each increases with higher contents
of sulfur in the fuel.

Carbon residue is a measure of the carbon remaining after all the volatile components of a fuel are vaporized in the
absence of air. Although it does not directly correlate with
engine deposits, carbon residue is considered to be an
approximation of carbon-depositing tendencies of a fuel.

Ash, non-burnable material in the fuel, is present in
two forms: abrasive solids and soluble metallic soaps. The
former contributes to engine deposits, plugging of the fuel
filter and fuel nozzle, and wear on the injector, injection
pump, pistons and rings. Soluble metallic soaps have little
effect on wear, but may contribute to engine deposits and
corrosion.

Stability is a measure of how well the quality of fuel
will be maintained in storage in contact with air and water.
If unstable components are present in the fuel, storage in air
can lead to formation of gums and sediments, which can cause
filter plugging, combustion chamber deposits, and gumming of
the injection system components, as well as engine wear.

Water and sediment can contaminate the fuel as a result
of poor handling and storage practices. Water has detrimental
effects on the components of the fuel injection system because
of its poor lubricating qualities.

Vegetable Oils

Swern reported (4):

Of the 250,000 species of plants known to botany
of which only perhaps 4,500 species were examined,
only 100 species are presently known to be oil-
bearing with sufficient oil content to warrant
commercial interest.

Of the 100, only about 22 vegetable oils are commercially

developed on a large scale today, and 12 of them constitute
more than 95% of the reported annual world vegetable oil
production. Swern lists twelve vegetable oil-bearing materials with their respective oil contents (Table 2).

Table 2.   Oil-Bearing Materials and Their Oil Content

Oil-Bearing Material	Oil Content (%)
1. Copra	65-68
2. Babassu	60-65
3. Sesame Seed	50-55
4. Palm Fruit	45-50
5. Palm Kernel	45-50
6. Groundnut (peanut)	45-50
7. Rapeseed	40-45
8. Sunflower Seed	35-45
9. Safflower Seed	30-35
10. Olive	25-30
11. Cottonseed	18-20
12. Soybean	18-20

Source: Swern(4)

     In general, vegetable oils are water-insoluble substances
of plant origin which consist predominantly of glyceryl esters
of fatty acids, so-called triglycerides. Structurally, a triglyceride is one molecule of glycerol esterified to three
molecules of long-chain monocarboxylic acids (fatty acids).
The resulting trigylceride or vegetable oil molecule has a
carbon chain which is much longer than the carbon chain of a
diesel fuel molecule. The most common fatty acids in vegetable oils are shown in Table 3. Fatty acids can be divided
into two classes, saturated and unsaturated. Each carbon
atom along the chain has the ability to hold two hydrogen
atoms. The fatty acid is saturated if all hydrogen atoms are
in place. If two adjacent carbons are missing hydrogen
atoms, the carbons bond doubly to one another, creating a
point of unsaturation. If there is more than one double
bond, the fatty acid is polyunsaturated. The relative amounts
of these fatty acids vary for the different vegetable oils
(Table 4).

     Different fatty acids affect the stability and melting
point of an oil. As the amount of unsaturation increases,
so does the relative rate of oxidation (Table 5). A vegetable oil that has a high degree of polyunsaturation would
tend to oxidize and polymerize sooner in the crankcase of an
engine than a vegetable oil which was highly saturated. On

Table 3. Common Fatty Acids in Edible Oils.

Name	Abbrev.	Formula	Class
Palmitic	C16	$CH_3-(CH_2)_{14} - COOH$	Saturated
Stearic	C18	$CH_3-(CH_2)_{16} - COOH$	Saturated
Oleic	C18:1	$CH_3-(CH_2)_7-CH=CH-(CH_2)_7 - COOH$	Unsaturated
Linoleic	C18:2	$CH_3-(CH_2)_4-CH=CH-CH_2-CH=CH-(CH_2)_7-COOH$	Polyunsaturated
Linolenic	C18:3	$CH_3-CH_2-CH=CH-CH_2-CH=CH-CH_2-CH=CH-(CH_2)_7-COOH$	Polyunsaturated

Table 4. Fatty Acid Composition of Common Vegetable Oils

ACID	CAPRYLIC	CAPRIC	LAURIC	MYRISTIC	PALMITIC	STEARIC	OLEIC	LINOLEIC	LINOLENIC	ERUCIC	SAPONIFICATION VALUE	IODINE VALUE
CARBON NO.	8	10	12	14	16	18	18	18	18	22		
DOUBLE BONDS	-	-	-	-	-	-	1	2	3	1		
COCONUT	2-6	3-7	44-54	14-19	6-10	1-4	9-19	1-3	-	-	243-255	14-24
PALM OIL	-	-	-	1-6	32-51	1-8	34-52	5-12	-	-	196-210	48-58
PEANUT	-	-	-	-	6-16	1-7	36-72	13-45	-	-	185-196	83-98
SAFFLOWER (HIGH OLEIC)	-	-	-	-	5	2	80	12	-	-	185-195	85-93
RAPESEED	-	-	-	0-2	2-5	0-2	13-30	10-25	5-10	20-50	168-183	94-106
COTTON SEED	-	-	-	0-2	17-19	1-4	13-44	33-58	0-2	-	189-199	103-115
RAPESEED (LOW ERUCIC)	-	-	-	-	3-4	1-2	54-58	18-??	6-12	0-5	188-195	110-115
SUNFLOWER	-	-	-	-	3-10	1-10	14-65	22-75	-	-	186-196	122-136
SOYA	-	-	-	-	7-12	2-6	19-30	48-58	4-10	-	188-195	124-136
SAFFLOWER	-	-	-	-	2-10	1-10	4-42	55-81	-	-	186-198	130-150

SOURCE: Adapted from Bacon, et. al. (5)

Table 5.   Relative Rates of Oxidation of Unsaturated
            Fatty Acids.

Fatty Acid	Relative Oxidation Rate
Stearic	0.6
Oleic	6
Linoleic	64
Linolenic	100

Source: Swern (4)

Table 6.   Melting Points of Triglycerides.

Name	Carbons	Melting Point (^{o}C)
Myristic	14:0	57.0
Palmitic	16:0	63.5
Stearic	18:0	73.1
Oleic	18:1	5.5
Linoleic	18:2	-13.1
Linolenic	18:3	-24.2

Source: Swern (4)

Table 7. Fuel Properties

	No. 2 Diesel Fuel	Sunflower Oil, Crude/Filtered
Density, kg/mm^3	847	921
Gross Heating Value, kJ/L	38,400	36,600
Cetane Rating	48	28
Viscosity, mm^2/s		
0°C	6.4	188
38°C	2.4	34
Pour Point, °C	-50	-9
Cloud Point, °C	-17	-7

Source: Kaufman and others (6)

the other hand, as the amount of saturation is increased, the melting point also increases (Table 6), so a vegetable oil with a high degree of saturation tends to be a solid at room temperature. A compromise between a low melting point and a high oxidation rate must be made when selecting a vegetable oil suitable for use in diesel engines. Unsaturation is a desirable property for maintenance of liquidity at low temperatures, but undesirable with respect to oxidative stability.

Iodine value, defined as the number of grams of iodine absorbed under standard conditions by 100 grams of fat, represents the degree of unsaturation in the fatty acid chain. Saponification number, defined as the number of milligrams of potassium hydroxide required to saponify one gram of fat, is a measure of the average molecular weight of fatty materials.

Differences in physical and chemical properties of vegetable oils, compared to diesel fuel, should be given some consideration before evaluating the use of vegetable oils as fuels for compression ignition engines. These differences vary in degree between the various types of vegetable oils. Table 7 lists some of the important fuel-related properties of crude filtered sunflower oil compared to No. 2 diesel fuel.

Ignition quality or the cetane number of vegetable oils has been reported to be from 28.3 (7) to 41.5 (3), compared to the minimum cetane number of 40 for No. 1-D and No. 2-D diesel fuels, specified by ASTM D975. In general, cetane numbers for the vegetable oils have been reported as being lower when compared to diesel fuel.

Another important difference appears in the heat of combustion, or calorific value. Vegetable oils have 5% to 18% less energy content than diesel fuel. The amount of decrease in energy content compared to diesel fuel is dependent on the type of the vegetable oil. An empirical equation can be used to calculate the gross heat of combustion based upon the saponification value and the iodine value (4):

Heat of Combustion (cal/g) = 11,380 - (iodine value) - 9.15 (saponification value).

Probably the greatest physical difference between the vegetable oils and diesel fuel is their viscosities. Vegetable oils are about 10 times more viscous at 40°C and about 30 times more viscous at 0°C. Viscosity is critically dependent on temperature, and the viscosity of vegetable oils is more seriously affected by temperature than that of

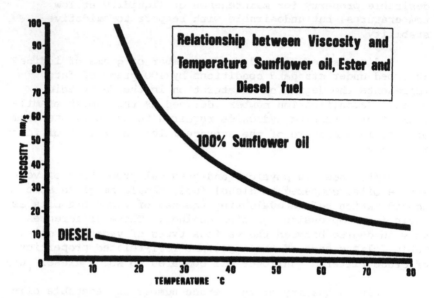

Fig. 1. Viscosity vs. Temperature for Sunflower Oil and Diesel Fuel. Reprinted by permission from Bruwer et al., "The Utilization of Sunflower Seed Oil as a Renewable Fuel for Diesel Engines," Agricultural Energy, Vol. 2, Publ. No. 4-81.

diesel fuels. Figure 1 shows the relationship between viscosity and temperature for sunflower oil and diesel fuel.

Other physical property differences include higher specific gravities along with higher flash, cloud, and pour points for vegetable oils. High specific gravities result in greater densities and weight per unit volume. A higher flash point reduces fire hazard. Higher cloud and pour points may become a limitation for the use of vegetable oils in colder climates.

One method of changing the physical properties of the vegetable oils to be more comparable with those of diesel fuels is to blend the vegetable oils with diesel fuel. Vegetable oils blend well with diesel fuel and do not show tendencies to separate or settle out. Table 8 shows some of the properties of vegetable oil/diesel fuel blends.

Mono-Esters

Another means of changing properties of a vegetable oil to become more comparable with those of diesel fuel is by converting them chemically to mono-esters. The process used to make this conversion involves reacting an alcohol with the vegetable oil in the presence of a catlyst. Three mono-ester molecules and a glycerol molecule are obtained from each tri-glyceride molecule. The glycerol, a by-product, is removed by water extraction. The final ester product is termed a methyl ester if methyl alcohol is used and ethyl ester if ethyl alcohol is used.

The viscosity of ester fuels was reported by Bruwer and coworkers (8) as being roughly of the same order as diesel fuel (Figure 2). Ester fuels also have distillation curves which are nearer to that of diesel fuel (8) (Figure 3).

Engine Tests

Short-Term Tests of Oilseed Fuels

Encouraging results have been obtained in short-term testing of modern diesel engines fueled with vegetable oils. Short term testing usually lasts only several minutes to several hours. The results of a number of postwar short-term engine tests on straight oilseed fuels was summarized by Quick (9). In summary, the short-term tests showed that power output, torque, and brake thermal efficiency on oilseed fuels equalled or were close to that of diesel fuel. Fuel consumption

Table 8. Vegetable Oil/Diesel Fuel Blend Properties

	Viscosity at 100°F, cSt	API Gravity at 60°F	Flash Point °F(°C)	Pour Point °F(°C)	Cetane Number	Gross Heat of Combustion, Btu/lb
Reference Diesel Fuel (railroad diesel)	3.46	32.0	159(71)	-58(-50)	44.3	19215
Peanut Oil						
25%	6.60	29.5		5(-15)	41.8	
50%	12.60	27.1	183(84)	16(-9)	40.5	
100%	39.51	22.7	622(328)	28(-2)	39.0	17045
Sunflower Oil						
25%	6.40	29.3		-4(20)	42.1	
50%	10.75	26.7	177(81)	-21(19)	40.8	
100%	33.45	21.9	608(320)	16(-9)	33.4	17010
Soy Oil						
25%	6.25	29.3		-13(-25)	43.6	
50%	11.28	26.7	179(82)	-21(-19)	41.9	
100%	32.31	21.9	597(314)	16(-9)	41.5	16770

Source: Southwest Research Institute (3).

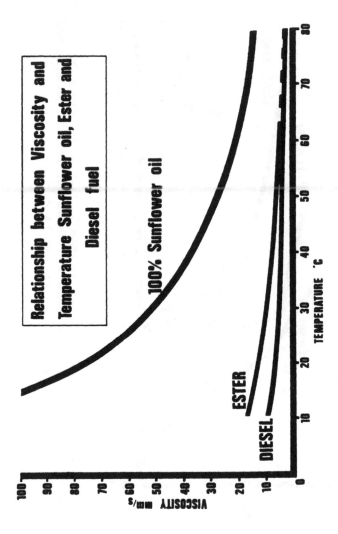

Fig. 2. The Relationship Between Viscosity Temperature: Sunflower Seed Oil, Ester, and Diesel Fuel. Reprinted by permission from Bruwer et al., "The Utilization of Sunflower Seed Oil as a Renewable Fuel for Diesel Engines," Agricultural Energy, Vol. 2, Publ. No. 4-81.

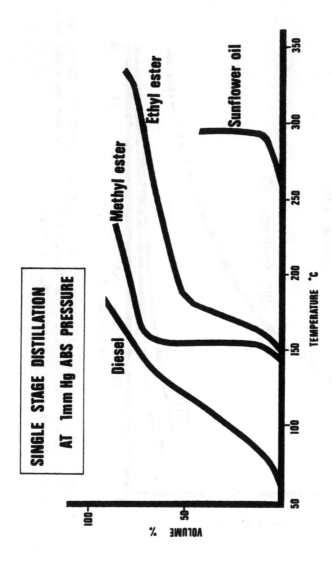

Fig. 3. Distillation Curves Showing Percentage Volumes Distilled at Various Temperatures. Reprinted by permission from Bruwer et al., "The Utilization of Sunflower Seed Oil as a Renewable Fuel for Diesel Engines," Agricultural Energy, Vol. 2, Publ. No. 4-81.

was invariably higher because of the lower calorific value of the vegetable oil.

Long-Term Tests of Oilseed Fuels

Although short-term combustion performance in an essentially unmodified diesel engine may be without incident, many researchers on vegetable oil fuels have found that the relatively poor thermal stability of vegetable oil leads to a buildup of deposits in the combustion chamber, especially injector nozzle coking, in long-term tests. The resultant degradation in fuel atomization and combustion efficiency leads to further problems such as piston ring sticking, crankcase oil dilution, and gelation of the lubrication oil resulting in engine failure.

Laboratory-Endurance Tests on Vegetable Oils in Direct-Injected Diesel Engines.

Early tests by deVedia (10) using a blend of 40% linseed oil/60% diesel fuel reported an increase in carbon deposits after 420 hours compared to 500 hours of operating on diesel fuel. The greatest increases were found on the cylinder head, intake valves, injection nozzles, pistons, and upper cylinder. Only a slight increase was found on the exhaust valves. The carbon deposits found using the blend, compared to diesel fuel, were harder and more tenaciously held to the engine parts. The engine used in these tests was a two cylinder, Lister diesel engine.

Aggarwal and coworkers (11) in tests on a single-cylinder engine of the same type concluded that more carbon deposits were formed using the vegetable oils as opposed to using diesel fuel. However, the deposits were not regarded as excessive or objectionable in character, perhaps because the duration of the test for each fuel was only 50 hours. Groundnut, polang, karanj, cottonseed, and rapeseed oils were tested.

A blend of 80% sunflower oil with 20% petrol was tested by Bruwer and coworkers (8) in a tractor equipped with a Perkins A4-248 engine to determine whether the lower viscosity of the blend could eliminate the injector coking problem encountered with neat sunflower oil. The engine was run on a dynamometer continuously for about 200 hours at 70% maximum power. At this point, the crankcase lubricating oil had polymerized to such an extent that the engine sump had to be removed in order to clean out the sludge. The injector nozzles were found to be coked up in the typical "trumpet" fashion around the nozzle holes. After refitting the sump and installing new injector nozzles, the test was resumed for 100 hours when the crankcase oil polymerized a second time. The engine was then dismantled and inspected. It was found that the piston rings were sticking, probably due to unburnt sunflower oil which had run down the cylinders. This in turn was probably caused by the coking of the injector nozzle holes which had prevented proper atomization of the fuel.

A blend of 60% sunflower oil with 40% illuminating paraffin was tested in a Ford 5000 engine, running for 100 hours at 70% maximum load (12). At the end of this period, the injectors were removed for inspection. It was found that the nozzles were coked up and the test was terminated.

McCutchen (13) durability-tested a turbocharged Caterpillar 3306 direct injection engine using 100% crude degummed soybean oil. A variable load cycle which was an approximation of field operation was used. Vegetable oil dripped from the exhaust stack during the idle portions of the variable load cycle. Fuel dilution in the lube oil was about 20%. At 110 hours, the engine developed a miss and could not carry full load. New injector nozzles corrected the problem. The removed nozzles were coated with a carbon deposit. Two nozzles had some plugged holes. At 130 hours, a piston seized and broke. The top side of the cylinder head under the valve cover was coated with a mixture of vegetable oil and lube oil which had polymerized. The oil pan and many oil passages contained a sludge of vegetable oil and lube oil which had started to saponify. A carbon layer had again started to build up on the fuel nozzles, primarily on the side nearest the exhaust valve.

Humke and Barsic (14) used a naturally aspirated, direct-injection diesel engine to evaluate the performance and emission characteristics of soybean oil, sunflower oil, and peanut oil. Injection nozzle deposits were observed for all tests. Carbon deposits formed on the injection nozzle tip exterior when using No. 2 diesel fuel were relatively soft and easily removed with a solvent-dampened cloth. Deposits formed with vegetable oils were harder. The two vegetable oils that produced the most difficult deposits to remove (crude soybean and peanut oil) also produced the greatest performance decreases and emissions increases. Deposits that formed on the nozzle tip exterior were believed to originate from thermal cracking of the fuel that boils from the nozzle sac late in the expansion stroke. Whether sufficient time is available for cracking to occur in an engine is not known. However, qualitative evaluation of deposit formation was consistent with fuel distillation results. Oils that thermally cracked at higher distillation fractions or higher temperatures formed less nozzle deposits and at slower rates than oils that cracked at lower distillation fractions or lower temperatures. Lubricating oil thickening, abnormal engine wear, and ring

sticking were not encountered during the test. However, none
of the vegetable oils was tested for more than 70 hours, and
the engine lubricating oil was changed for each different fuel.
Also, injection nozzles were cleaned every 10 hours.

Bacon and coworkers (15) reported that tests on a single-
cylinder, direct-injection, naturally aspirated Perkins diesel
engine showed deposits formed on the injector nozzle after
only a few hours of testing using corn, soya, and sunflower
oils. These carbon deposits formed after only two hours of
running time on corn and soya oils and after nine hours of
sunflower oil.

Bacon and coworkers (5) also evaluated the polymerization
effects of sunflower oil, coconut oil, and hardened tallow in
a single cylinder version of a typical direct injection diesel
engine (Perkins 4.236). The engine was run continuously at
part load, mid speed for two hours. These conditions were
established by previous tests to produce the most rapid nozzle
coking tendencies. The vegetable oils were chosen because
they represented a range from fully saturated to highly unsaturated. The saturated fuels (coconut oil and hardened
tallow) produced much less injection nozzle coking than did
the unsaturated fuels (sunflower oil). These tests serve
only as evidence for the chemical hypothesis of nozzle coking
but do not recommend highly saturated oils on a practical
basis since they are solid at normal ambient temperatures.

Quick and coworkers (16) also used the iodine value as a
criterion for the comparison of vegetable oils for coking.
They selected linseed oil as the candidate "worst" oil to
obtain baseline data as a standard comparison for future
work. Their assumption was that if there was some additive
or fuel modification anticipated to reduce injector fouling
it would become most readily apparent if it was evaluated
first against or blended with linseed oil. A duration test
was conducted at constant speed on a Lombardini 720 natural-
ly-aspirated, direct-injection, air-cooled, single cylinder,
diesel engine. The engine had to be shut down after only
16 hours of operation when power had declined to around 67%
of the original value. The engine was subsequently examined
and it was noted that the area around the injector was completely fouled. The four injector holes were delineated
by funnels of carbonaceous materials extending as far as 10 mm
away from the injector tip. Some deposits were beginning to
form in the piston combustion space and around the valves and
ports, although the rings were still free and the oil was not
visually severely contaminated. A blend of 10% 1-butanol and
90% linseed oil was evaluated to see whether it would have any
decoking capability. The Lombardini engine was also run on

heated (85°C) and cold linseed oil to see if any viscosity effect was apparent in extended operation. The results of these tests were negative. There was no significant increase in hours logged before power diminished when operating on heated linseed oil or on the 1-butanol blend.

Galloway (17) used crude peanut oil for the purpose of long-term effect analysis of vegetable oil fuels. Two identical Lister SR 2 twin cylinder, air-cooled, direct-injection diesel engines were run at full load, 5 kW, for 1000 hours. One engine was operated on diesel fuel and the other engine was operated on crude peanut oil. The high viscosity of the peanut oil resulted in no leak back being evident at the injectors of the peanut oil fueled engine. It was also found that the static pressure on vegetable oil was very much greater than for conventional diesel fuel. This effect reduced the seating pressure of the injector needle valve and created a situation where post injection dribble could occur. Under the conditions present with high viscosity vegetable oil, nozzle temperature increased considerably due to negligible leak back and inhibited pressure relief. Polymers developed in the clearance area between the nozzle valve and its barrel in the injector. Consequently, carbon trumpets formed in 48 hours of continuous running as a result of at least two conditions - high nozzle temperature and thermal polymerization of some of the fuel trapped in the annular chamber in the nozzle tip.

The oil change interval was 250 hours. The viscosity of the lubricating oil in the research test engine fueled by peanut oil increased abnormally during the normal 250 hours on an oil change indicating crankcase oil dilution. For the period from 275-525 hours of engine running, three oil samples were taken from each engine. An atomic absorption spectographic analysis of the oil samples indicated that wear metals most important with respect to engine deterioration for the peanut-fueled engine were significantly above those for the diesel-fueled engine. It seems reasonable to presume that the metals highest for the peanut-fueled engine were both abrasively and chemically released. The injectors of the engine on peanut oil were then modified to provide significant back leakage. It was presumed that this modification would reduce the thermal polymerization of fuel lodged between the valve stem and its housing wall, provide lubrication, and partially reduce the pressure relief problem. Trumpet growth during the next 250 hours of running was considerably reduced and combustion greatly improved. Oil sample results from the peanut fueled engine for the period of 275-525 hours compared with the period of 577-826 hours indicated that the rate of increase in wear metals in the second test was considerably lower than the first test.

Peters and Ziemke (18) used blends of two-thirds No. 2
diesel fuel, one-third degummed soybean oil (2:1) and one-
half No. 2 diesel fuel, one-half degummed soybean oil (1:1) to
fuel separate 200-hour test runs on a Deere & Company 6404 TE
turbocharged direct injection engine. The recommended crankcase oil change interval for this engine is 100 hours, but
an extended change interval of hours was used in order to gain
data on crankcase oil contamination and carbon build-up on
fuel injectors. Brake thermal efficiency and maximum horsepower measurements were obtained at intervals. No significant
change in crankcase oil viscosity was noted during the 200
hour test run using one-third soybean oil and two-thirds No. 2
diesel fuel. The change in crankcase oil viscosity was more
pronounced when a blend of half soybean oil and half No. 2
diesel was used. Carbon build-up on fuel injectors was more
pronounced when the 1:1 fuel blend was used than with the 2:1
fuel blend. In general, the carbon build up in 200 hours
using the 2:1 fuel blend was soft and could be removed easily.
Hard carbon buildup was noted when the 1:1 fuel blend was
used. It was concluded that the blend of the one-third degummed soybean oil and two-thirds No. 2 diesel fuel is a
viable fuel for turbocharged direct injection engines. There
were insignificant changes in corrected brake thermal
efficiency and corrected maximum horsepower during the 200-
hour test. A blend of half degummed soybean oil and half No.
2 diesel fuel is a marginal fuel. Both corrected brake thermal efficiency and corrected maximum horsepower declined
during the 200-hour test.

Two endurance tests were completed on a direct injected,
turbocharged and intercooled diesel engine in the Allis
Chalmer Engine Division at Harvey, Illinois (19). The two
tests included a 500 hour baseline test with No. 2 diesel fuel
and a 600 hour test with a 50/50 blend of alkali refined sunflower oil with No. 2 diesel fuel. A continuous test cycle
of three minutes at high idle and 10 minutes at peak torque
was used. During the baseline test there were no significant
problems with engine operation. However, problems were
experienced while operating on the blended fuel. While using
the blended fuel: 1) there was abnormal carbon buildup on the
injection nozzle tip; 2) the nozzle opening pressure dropped
10% to 15%; 3) the injection nozzle needles stuck; 4) there
was excessive carbon buildup on the intake ports; 5) operating
problems of the turbocharger were observed, and 6) piston
ring sticking problems were experienced.

Baranescu and Lusco (20) evaluated the performance and
durability of direct-injection turbocharged diesel engines

using sunflower oil and blends thereof. Alkali refined sunflower oil and three different blends of sunflower oil and
diesel fuel were comparatively tested against No. 2 diesel
fuel for physical and chemical characteristics, fuel injection
system performance, short-term engine performance, propensity
for nozzle deposits buildup, limited durability operation, and
low temperature starting capability. They concluded that the
use of sunflower oil blended with diesel fuel in a direct
injection engine brought about modifications in the fuel
injection process characteristic to injection of viscous fuel,
namely; increase in injection pressure, longer injection duration, delayed needle closing, early needle opening. All
these effects were conducive to longer combustion duration.
The modified injection characteristics and the blend properties generated a peculiar spray pattern with larger drops
and   fuel overpenetration. The subsequent combustion followed a different mechanism of energy release.

The increased propensity to leave residues behind made
the blends with sunflower oil prone to deposit buildups after
only short periods of operation. The tests showed that nozzle
deposits tend to be blown off to some extent during operation,
while piston and liner deposits were more stable. The deposit
buildup occurred even at high load and speed. Low speed and
load operation increased the tendency to buildup deposits by
addition of unburned fuel. Durability testing of 200 hours
was insufficient for revealing long-term performance of sunflower oil blends. Engine shutdown for longer periods of time
accelerated gum formation where the fuel contacts the bare
metal, which is likely to further impair engine and injection
system function. Finally, cold temperature operation was a
very critical issue related to sunflower oil and blends
thereof. High viscosity caused fuel system problems, failed
starting, unacceptable emission levels, and injection pump
failures due to lack of lubrication. In general, Baranescu
and Lusco's work showed that vegetable oils are candidates
for alternative fuels due to their similarities with conventional fuels. However, due to their differences, vegetable
oils and their blends are not yet a viable practical solution
since they affect the performance, the durability and the
serviceability of the engines, especially the more fuel-
sensitive direct-injection engines.

Fort and coworkers (21) tested experimental fuels of
cottonseed oil, transesterified cottonseed oil (methyl ester)
and No. 2 diesel fuel in a turbocharged, open chamber diesel
engine. Emphasis was placed on durability effects but information on basic performance and emissions was obtained.
The durability cycle selected followed a proposal of the
Engine Manufacturers' Association (EMA) for engine screening

work with alternative fuels. At the time of the report, two
fuels, 50% cottonseed oil/50% No. 2 diesel fuel and 100%
transesterified cottonseed oil had been evaluated. After 183
hours of the cyclic test on the 50/50 blend, the engine was
noisy at idle and had excessive blowby. Borescope inspection
of the cylinder revealed scoring in Nos. 1 and 5. The scoring
was serious enough to end the test at that point. The engine
was disassembled, inspected, rated for deposits, and photo-
graphed. By contrast, the test ran the full 200 hours on
transesterified cottonseed oil with no noticeable change in
engine power or exhaust temperature and without any detect-
able problems. When the engine was disassembled the general
appearance was about the same as if diesel fuel had been
used. However, a ring groove deposit effect was noted which
was more pronounced than with the cottonseed oil/diesel fuel
blend and was the most serious potential problem noted for
transesterified cottonseed oil.

   Field Endurance Tests on Vegetable Oils in Direct-
Injected Diesel Engines.   Tests totalling 1382 engine meter
hours were conducted on a Ford 5000 diesel tractor operating
on 100% sunflower oil and a 20% sunflower oil/80% diesel fuel
blend (8). The first 100 hours was a maximum power dynamometer test using 100% sunflower oil with no noticeably
adverse effects reported. The tractor was then used in field
tests running on the 20:80 blend. At the completion of an
additional 1004 engine meter hours of trouble-free operation,
an 8% power loss was measured at the power take-off. The
injector nozzles were then replaced and the injection pump
recalibrated to specification. This reduced the power loss to
only 4%. Following the field test, the tractor was coupled
to a power take-off dynamometer and operated at a continuous
load of 70% maximum power using the same 20:80 blend. The
engine ran 24 hours a day for a total of 278 engine meter
hours when the exhaust smoke increased noticeably. The injector nozzles were removed to find that they had started to
carbon up around the orifices. The rest of the engine was
then disassembled to find that deposits in the combustion
chamber, cylinders and piston ring grooves were indistinguish-
able from those when operating normally on diesel fuel.
Another make of tractor operating at the same 70% maximum power load, with a blend of 80% sunflower oil/20% gasoline for
about 300 hours, however, did exhibit piston ring sticking and
deposits forming on the injector tip.

   Van der Walt and Hugo (22) used four different makes of
tractors to determine whether the encouraging results (8) with
a Ford 5000 tractor running 20% degummed sunflower oil/80%
diesel fuel blend would also be applicable to other makes of
tractors. The tractors tested were a Massey Ferguson 285,

a Fiat 880, an International Harvester 844S and a John Deere
2030. All these tractors have direct-injected diesel engines.
They were all used in field test conditions. It is suspected
that, initially, poor mixing of sunflower oil and diesel oil
caused the tractors to run on neat sunflower oil for short
periods of time. This could possibly have had a detrimental
effect on the engines before special efforts were made to mix
the fuel properly.

The Massey Ferguson 285 tractor started to smoke excess-
ively after 260 hours of field operation on the mixture, and
cold starting was getting difficult. New injectors were fitted and the test continued. After 392 hours in the field,
power was diminished by 4.7% from the initial value. Lubricating oil analysis showed normal wear rates for engine
components.

After 576 hours, the Fiat 880 tractor operator complained
about low power. It was found that the compression on two
cylinders was very low. Further investigation revealed
severely coked injectors and combustion chambers as well as
sticking piston rings. Lubricating oil analysis showed high
piston, liner, and bearing wear.

Injectors were coked after 361 hours on the International
Harvester 844S. After new injectors were fitted and the
injection pump was recalibrated, power was still down by 7.4%
from the initial value. Lubricating oil analysis showed high
ring, liner, piston, and bearing wear. The tractor is still
operating in the field, having completed 539 hours.

The John Deere 2030 completed 396 hours without apparent
problems. A power check revealed a 5.7% loss in power at
that stage. Lubricating oil analysis indicated normal wear
rates for the engine components.

Not one of the four direct-injection engines could complete much more than 400 hours on the 20% sunflower oil, 80%
diesel mixture without coking injector tips or power loss.
These results, however, could have been affected by initial
operation on neat sunflower oil.

Sims and coworkers (23) conducted field tests on three
tractors using alkali-refined rapeseed oil fuels. The tractors tested were a Leyland tractor with a Perkins 3 cylinder
diesel engine, a Massey Ferguson 165, and a Ford 5000. The
Leyland and the Massey Ferguson used 99% rapeseed oil and 1%
two stroke oil. During the early part of the trials there
was concern that gum deposits would buildup in the injection
system, so the Ford 5000 used a blend of 19% dieselene, 1%

Wynns diesel fuel conditioner, and 80% rapeseed oil. No
such deposits were experienced. Initially, 100% rapeseed oil
did not provide adequate lubrication to rotors in rotary
distribution type fuel injection pumps and seizure of rotor
shafts occurred within only a few hours. Preheating of fuel
to 10°C reduced its viscosity but did not prevent these pump
failures. It appeared that the film strength of rapeseed oil
was insufficient to withstand the pressures encountered in
distributor type pumps. The addition of 1% good quality Shell
Super two-stroke lubricating oil or 19% dieselene blend
appears to have overcome this problem. The higher viscosity
of rapeseed oil affected atomization of the fuel resulting
in increased carbon deposits on the injector tips during
cold running under light load. These deposits were of a soft
fluffy nature and did not affect cold starting or maximum
engine performance. No excessive smoke emissions were experienced with these deposits, and they quickly disappeared
after a short period of operating under heavy load conditions.
Due to the very high viscosity of the rapeseed oil, only half
the normal quantity of leak back past the injectors occurred,
but no needle sticking, scoring or nozzle overheating occurred. As this bypass fuel cools the injectors, injector nozzle overheating could be expected on some engines with the
use of this fuel. Engine oil compatibility with the rapeseed
oil was found to be satisfactory when the oil change periods
were carried out in accordance with the engine manufacturer's
recommendations. No sludge deposits or insoluble gels were
experienced in the lubricating oil. The oil change periods
did not exceed 200 hours and the oil was Shell Agroma 15W/30.

The Leyland engine was dismantled after 480 hours, revealing carbon deposits similar in quantity to those produced
by dieselene, but softer. No gum deposits were found, and
fuel injection equipment appeared as new. No significant
wear was found on pistons, rings, and other moving parts.
Valve faces were in satisfactory condition.

M. Worgetter (24) field tested a 43 kW four cylinder
direct injection Stery diesel tractor fueled with a mixture
of 50% (v/v) rapeseed oil and diesel fuel. The rapeseed oil
was of food quality. The only modification to the engine was
to substitute an in-line injection pump for the factory
standard distributor injection pump. The tractor was engaged
387 hours to perform various farm tasks. At the end of the
test the cylinder head had a lacquer-like coat of 0.5 to 1 mm.
The exhaust ducts were narrowed with soot and carbon residues,
the cross sectional area being reduced by 30%. The inlet pipes
were free. The inlet valves were coated with carbon residues
on the stems (thickness appr. 3 mm); the valve seats were in
good order. The exhaust valves had a thinner coating but the

seats were coated, too. The piston crown was covered by a thin
layer of carbon residues, and the combustion chamber was soot-
ed. The piston area above the first piston ring was covered
with very hard carbon deposits. The first piston rings were
firm, the second rings could be moved partly, the third rings
were free. Carbonaceous matter had also formed in the slots
of the scraper rings. The cylinder liners showed hard carbon
deposits above the first piston ring and this coating partially extended to the second piston ring. All the bearings of
the engine were in good working order. The injection nozzles
had heavy carbon residues ("trumpets") of a length of 1 to
2 mm. The injection pump was examined by the manufacturer.
The delivery performance of the particular pump cylinders
was not uniform. No gumming and corroding effects could be
observed. An optical check of the cam and the roller surface
did not show any particular wear.

Walter and coworkers (25) tested twelve unmodified tractors in the field on blends of sunflower oil and diesel fuel.
John Deere, Allis Chalmers, and J.I. Case tractors ranging
in power from 89.5 kW to 186.4 kW (120 hp to 250 hp) were
used in the program. Six tractors ran on a 25:75 sunflower
oil/diesel fuel blend. The remaining tractors operated on
a 50% blend. The sunflower oil was alkali refined and
winterized. Nearly 7000 hours were accumulated on the 12
tractors during the 1981 crop season. Cooperating farmers
started and operated their tractors at temperatures ranging
from -15°C (5°F) to over 40°C (104°F). Fuel filtration
problems were not apparent and engine performance was sat-
isfactory. Following the end of fall work, the tractor
engines were disassembled to check for abnormal wear and
engine deposits. Inspection revealed high carbon and
varnish-like deposits on several engine components. Piston
ring grooves and ring lands on all engines contained excessive deposits. Use of pistons with a flat ring design
running on the 50 percent blends led to ring sticking, a few
broken oil rings, and scored cylinder sleeves. Engines with
keystone (tapered) piston rings did not have stuck rings,
but the level of carbon and varnish appeared greater than
would normally be expected. The underside of the engine
head including the valve heads and injector tips appeared
relatively free of excess carbon. The load characteristics
of the tractor appeared to have as large an influence on
engine deposits as the percentage of sunflower oil in the
fuel blend. Light loads appeared to increase deposits.
Nearly all tractors had excessive carbon on intake valve
stems and intake ports. These deposits could lead to valve
sticking and burned valves. Engine wear appeared normal
except in the case where sticking piston rings led to scored
sleeves. Engine bearing wear appeared normal in all engines.

Laboratory Endurance Tests on Vegetable Oils in
Indirect Injected Diesel Engines. Fang (26) reported heavy
carbon deposits were found in the combustion chamber of a
single cylinder, indirect-injection, naturally aspirated,
Atlas-Lanova diesel engine during tests using castor oil and
two grades of soybean oil. After 39 hours of running on
diesel fuel, followed by 22 hours with the various vegetable
oils, the engine started to lose power and increase smoke output. The engine was disassembled to find that carbon deposits
in the combustion chamber had partially closed the air intake
valve opening. Some gumming was also found on the intake
valve stem. Careful examination of the fuel nozzle, injection
pump, valve, and valve seat showed no irregularities in these
parts. After the engine was cleaned and reassembled it performed as well as at the start of the tests.

McCutchen (13) durability-tested naturally aspirated and
turbocharged prechamber versions of a Caterpillar 3306 diesel
engine using crude degummed soybean oil. A variable load
cycle, which was an approximation of field operation, was
used. Naturally aspirated prechamber operation resulted in
pistons and rings cleaner than would have been expected from
burning diesel fuel, in sharp contrast with similar tests on
the direct-injected model 3306, previously described. A
slight buildup of carbon on the fuel nozzle tip seemed to
stabilize. The inside of the prechamber was coated with
a soft carbon that seemed to buildup, flake off and rebuild.
The comparable diesel prechamber had a thin hard carbon
coating that also flaked off and rebuilt. After 200 hours
of operation, piston, ring, and liner wear was too small to
measure. There was no fuel dilution of the lube oil. A
mixture of 30% soy oil/70% diesel fuel and 100% soy oil both
gave excellent results. Turbocharged operation with a
mixture of 30% crude degummed soy oil/70% diesel fuel caused
problems. The wear rates at lower ratings were excessive.
However, the wear rates at lower ratings approached those
expected in some fuel short areas where petroleum sulfur
levels exceed 1%. The 3306 engine has a single inlet and
exhaust valve with the prechamber offset and tilted away from
the valves. The greatest cylinder liner wear plus a tendency
for scuffing at high ratings occurred on the side nearest
the prechamber and opposite the valves. McCutchen's theory
is that the vegetable oil burns more slowly than diesel fuel
and that partially burned oil is quenched on the cylinder
liner in the highest wear areas. The fuel/oil/air mixture
directed across the cylinder from the prechamber has a longer
time for combustion, and so less partially-burned carbon is
deposited on the liner. Caterpillar will permit the use of
30% mixes of soybean, sunflower, peanut and rapeseed oils with
diesel fuel in precombustion chamber engines in construction

machinery operating in Brazil. Castor oil is excluded from
consideration because it will not stay mixed with fuel. This
was done after considerable testing and after consideration
of the engine ratings and duty cycles of the engines involved.
If successful, the approval may be extended to other areas
of the world.

A Deutz F3L912W indirect injection engine was installed
in a standard Deutz tractor and subjected to the manufacturer's cyclic load for 600 hours duration, running on neat
sunflower oil in the laboratory, coupled to a PTO dynamometer
(22). On completion of the test the engine was dismantled
and inspected in detail by an expert from the manufacturer.
It was reported that there were no signs indicating that
sunflower oil was not a suitable fuel for that engine. This
test is being repeated. In another test, the same type of
engine was subjected to a 70% maximum power load, running
continuously for almost five months. Normally, the engine
was only stopped for routine maintenance and oil sampling.
Apart from a burnt exhaust valve on one cylinder after 1,000
hours, and again a burnt exhaust valve on another cylinder
after a total of 2,000 hours, the engine ran without problems.
After 2,300 hours, the engine unfortunately overheated due
to a fault in the fresh air supply, and the test is being
repeated.

Peterson and coworkers (27) ran two single-cylinder,
water-cooled, indirect injection Yanmare diesel engines
in an endurance test. One engine was operated on diesel fuel
for 840 hours. The other engine was operated on 100%
safflower oil for 830 hours. Both engines were connected to
electric generators. The generator load was 85% of the
rated engine capacity and was cycled on and off with a 15
minute cycle time. At the completion of the test, both
engines were completely disassembled to compare deposits and
wear. In general, results showed the safflower fueled engine
to have a wear rate about twice the rate of the diesel-
fueled engine but at a rate such that all components were
still within acceptable limits. The safflower-fueled
engine showed more carbon in the exhaust ports, combustion
chamber, on the piston rings, and on the injector nozzles.

Borgelt and Harris (28) tested three Onan 5 kW (6.7 hp)
single cylinder, air cooled, prechamber engines for 1000
hour endurance runs at 50 to 55% of rated power. Each engine
operated on a different fuel. One engine used 100% No. 2
diesel as a baseline. The other two engines used a 25/75
soybean oil diesel fuel blend and a 50/50 soybean oil/diesel
fuel blend respectively. No significant difference between
engines occurred regarding ring, cylinder, bearings, and

journal wear. Some increase in carbon was observed on the
engine heads as the percent of soybean oil increased. The
intake valve stem and port of the 50/50 engine did show
considerable buildup of carbon over the other two engines.
Pistons and rings showed no evidence of carbon deposits,
sticking parts, or other detrimental effects. Oil analysis
for all three engines indicated no abnormal wear. Oil
change intervals were maintained at 100 hours. Fuel filtration was a problem for the 50% engine. Eight filter
changes were made during the 1000 hours of the test. Upon
disassembly of the injectors, both the 25/75 and 50/50 engine
injectors displayed varnish and gum deposits.

Laboratory Endurance Tests of Mono-Esters. McCutchen
(13) durability-tested a turbocharged Caterpillar 3306
direct injection engine using methyl ester of rapeseed oil.
A variable load cycle, which was an approximation of field
operation was used. The methyl ester operated successfully
for 250 hours. At the end of the test the sides of the
pistons were clean. Some oil dilution was observed but
without any increase in viscosity or other problems. The
piston crowns and exhaust valves were coated with an iron
oxide which seemed to come from the fuel. A free fatty
acid level of 8.7% was blamed for the iron which seemed to
originate in the barrels used to transport and store the
oil. The fuel nozzles had deposits similar to those in
operation with straight soy oil but no performance deterioration was observed.

Bacon and coworkers (5) evaluated the nozzle coking
problem of modified vegetable oil in a single cylinder
version of a typical direct injection diesel engine (Perkins
4.236). The engine was run continuously at part-load,
mid-speed for two hours. These conditions were chosen by
previous testing to produce the most rapid nozzle coking
tendencies. The triglyceride variants were ethyl esters
of sunflower oil (Iodine value = 132), ethyl oleate (Iodine
value = 82), and methyl stearate (Iodine value = 1), rep-
resenting a range of saturation levels. The injector nozzles
of the saturated fuels had much less injector nozzle coking
than the unsaturated fuel. The results also indicated that
the injector coking problem was less severe with mono-esters
than with unmodified vegetable oils.

Conclusions

Modern diesel engines have been designed to operate on
standard diesel fuel. Diesel engines require fuels which
self-ignite readily at compression ignition temperatures.
Vegetable oils have surprisingly good fuel properties for

application in compression ignition engines. They are
safe to store and handle and they are the only renewable
fuel which can currently power compression ignition engines
with some degree of success. The short-term performance
in an unmodified diesel engine is satisfactory. However,
there are problems associated with the long-term use of
vegetable oil fuels, explained by physical and chemical
properties of vegetable oils that differ considerably from
those of No. 2 diesel fuel. The major technical problems
with vegetable oils are their high viscosity, unsaturation,
very low volatility, and tendency to form residues on
combustion.

Long-term tests indicate that vegetable oils and
blends of these oils with diesel fuel can have seriously
adverse effects on direct injection (DI) engines, the
most common type of diesel engine. Virtually all of the
tests of vegetable oils and blends in DI engines have
produced rather heavy deposits on the injector nozzles.
In addition, most tests resulted in substantial deposits of
carbon or varnish on other engine components including
valves, cylinders, pistons and rings. These deposits
are believed to be caused by the high viscosity of the vegetable oils which leads to poor atomization of the fuel and
inefficient combustion. Residues of partially combusted
fuel remain in the engine to form deposits or to contaminate
the lubricating oil. Some of the tests have indicated
serious contamination of the lubricating oil with unburned,
polymerized fuel and/or metals. There are indications that
the problems of engine deposits and lubricating oil contamination are more serious with vegetable oils that have
high iodine values. The most frequently tested vegetable
oils, sunflower oil and soybean oil, have relatively high
iodine values. Coconut oil has an iodine value of 20 compared to 130 for soybean oil which may account for the
lack of reported problems with engine deposits in the case
of coconut oil. However, the problem with coconut oil is
that it is a solid at normal ambient conditions.

Some research indicates that straight vegetable oils
or blends of these oils with diesel fuel can be used in
indirect injection (IDI) engines without carbon deposits or
oil contamination problems.

A possible solution to the problems of long-term
operation of DI engines on vegetable oils may be achieved by
the chemical process of transesterification. The esters
have low viscosities, due to removal of part of the molecule
in the form of glycerol. Several tests indicate that ester
fuels can be used in DI engines without the carbon deposits

which accumulate when straight vegetable oils and blends are used in these engines.

Overall, more endurance testing is needed before any firm conclusions can be drawn about the use of vegetable oils as fuels in any type of diesel engine. At this writing, their use in the United States cannot be considered either economically or technically feasible.

References

(1) R.W. Nitske, and C.M. Wilson. "Rudolf Diesel, Pioneer of the Age of Power," 1st Edition, University of Oklahoma Press, 1965.

(2) E.F. Obert. Internal Combustion Engines and Air Pollution, Harper and Row Publishers, New York, NY 1973.

(3) "Emergency Fuels Utilization Guidebook," DOE/CS. 54269-01 Southwest Research Institute, San Antonio, TX. August, 1980.

(4) D. Swern (Editor). Bailey's Industrial Oil and Fat Products. Vol. 1, 4th Ed. (John Wiley and Sons, New York, 1979).

(5) D.M. Bacon, F. Brear, I.D. Moncrieff, and K.L. Walker, "The Use of Vegetable Oils in Straight and Modified Forms as Diesel Engine Fuels," In Beyond the Energy Crisis, Vol. III. Fazzolare and Smith, Eds. (Pergamon Press, Oxford, England. 1981) pp. 1525-1533.

(6) K.R. Kaufman, M. Ziejewski, M. Marohl, H. Kucera, and A.E. Jones. 1981. Performance of diesel oil and sunflower oil mixtures in diesel farm tractors. Paper #81-1054. American Society of Agricultural Engineers, St. Joseph, MI.

(7) P. Ramdeen, L.F. Backer, K.R. Kaufman, H.L. Kucera, and C.W. Moilanen. "Some Physio-Chemical Tests of Sunflower Oil and No. 2 Diesel Oil Fuels," ASAE Paper No. NCR 81-009. (American Society of Agricultural Engineers, St. Joseph, MI, 1981).

(8) J.J. Bruwer, B.v.d. Boshoff, F.J.C. Hugo, J. Fuls, C. Hawkins, A.N. v.d. Walt, A. Engelbrecht, and L.M. dePleassis. The Utilization of Sunflower Seed Oil as a Renewable Fuel for Diesel Engines. Agricultural Energy, Vol. 2 (American Society of Agricultural Engineers, Publ. No. 4-81, St. Joseph, MI 1981) pp. 385-390.

(9) G.R. Quick "Developments in Use of Vegetable Oils as
     Fuel for Diesel Engines", ASAE Paper #80-1525.
     (American Society of Agricultural Engineers, St. Joseph,
     MI, 1980)

(10) M.R. deVedia. "Vegetable Oils as Diesel Fuels,"
     Diesel Power and Diesel Transportation. Vol. II, No. 12.
     December, 1944.

(11) J.S. Aggarwal, H.D. Chowdhury, S.N. Mukherji, and
     L.C. Bermah. "Indian Vegetable Oils as Fuels for
     Diesel Engines," CSIR Bulletin, No. 19. New Delhi,1952.

(12) J.J. Bruwer, F.J.C. Hugo, and C.S. Hawkins. Sunflower
     Oil and Esters as Tractor Fuel. Paper No. 81037.
     National Conference on Fuels from Crops, (Society of
     Automotive Engineers - Australasia, Victoria, Australia,
     1981.)

(13) R. McCutchen "Vegetable Oil as a Diesel Fuel - Soybean
     Oil." In Beyond the Energy Crisis. Vol. III.
     Fazzolare and Smith, Eds. (Pergamon Press, Oxford,
     England, 1981) pp. 1679-1686.

(14) A.L. Humke, and N.J. Barsic. "Performance and Emissions
     Characteristics of a Naturally Aspirated Diesel Engine
     with Vegetable Oil Fuels (Part 2)," SAE Paper #810955,
     (Society of Automotive Engineers, Warrendale, PA.
     Sept., 1981).

(15) D.M. Bacon, N. Bacon, I.D. Moncrieff, and K.L. Walker.
     "The Effects of Biomass Fuels on Diesel Engine Combustion
     Performance", Perkins Engines Group Limited.
     Peterborough, England, 1980.

(16) G.R. Quick, B.T. Wilson, and P.J. Woodmore, "Develop-
     ments in Use of Vegetable Oils as a Fuel for Diesel
     Engines. Part II. Studies on Injector Coking. Paper
     No. 81020. National Conference on Fuel from Crops,
     (Society of Automotive Engineers - Australasia, Victoria,
     Australia, 1981.)

(17) D.J. Galloway "Vegetable Oils and Animal Fats as
     Diesel Fuel," Paper No. 81021 National Conference on
     Fuel from Crops, (Society of Automotive Engineers -
     Australasia, Victoria, Australia, 1981.)

(18) J. Peters, and M. Ziemke. "Second Interim Report Test
     of Plant Oil (soybean) as a Diesel Fuel," UAH Report
     No. 279. Kenneth E. Johnson Environmental & Energy

Center.   The University of Alabama in Huntsville, AL
1981.

(19) M. Ziejewski, and K.R. Kaufman.   Endurance test of a
sunflower oil/diesel fuel blend.   SAE Paper #820257
(Society of Automotive Engineers, Warrendale, PA, 1982).

(20) A. Baranescu and J. Lusco.   "Sunflower Oil as a Fuel
Extender in Direct Injection Turbocharged Diesel
Engines", SAE Paper #820260.   (Society of Automotive
Engineers, Warrendale, PA, 1982).

(21) E.F. Fort, P.N. Blumberg, H.E. Staph, and J.J. Staudt.
"Evaluation of Cottonseed Oils as Diesel Fuel,"   SAE
Paper #820317.   (Society of Automotive Engineers,
Warrendale, PA, 1982).

(22) A.N. Van der Walt and F.J.C. Hugo.   "Diesel Engine Tests
With Sunflower Oil as an Alternate Fuel,"   In Beyond
the Energy Crisis, Vol. III.   Fazzolare and Smith, Eds.
(Pergamon Press, Oxford, England, 1981).   pp. 1927-1933.

(23) R.E.H. Sims, R.R. Raine, and R.J. McLeod.   "Rape Seed
Oil as a Fuel for Diesel Engines," Paper No. 81022.
National Conference on Fuels from Crops, (Society of
Automotive Engineers, Australasia, Victoria, Australia,
1981).

(24) M. Worgetter, "Results of a Long Term Engine Test
Based on Rape Seed Oil Fuel,"   In Beyond the Energy
Crisis, Vol. III.   Fazzolare and Smith, Eds. (Pergamon
Press, Oxford, England.   1981)   pp. 1525-1533.

(25) J.R. Walter, P. Aakre, and J.Derry.   The North Dakota
"Flower Power" Project.   North Dakota Farm Research,
Vol. 39, No. 6, North Dakota Agricultural Experiment
Station, Fargo, ND.   May-June, 1982.

(26) K.S. Fang, "Vegetable Oils as Diesel Fuels for China,"
M.S. Thesis.   University of Nebraska, Lincoln, NE, 1949.

(27) C.L. Peterson, G.L. Wagner, and D.L. Auld.   "Performance
Testing of Vegetable Oil Substitutes for Diesel Fuel,"
ASAE Paper #81-3578, (American Society of Agricultural
Engineers, St. Joseph, MI, 1981)

(28) S.C. Borgelt and F.D. Harris.   "Effects of Soybean Oil-
Diesel Fuel Mixtures in Small Pre-Combustion Chamber
Engines",   ASAE Paper #MC82-144.   (American Society
of Agricultural Engineers, St. Joseph, MI, 1982).

William Lockeretz

10. Seed Oils as Diesel Fuel: Economics of Centralized and On-Farm Extraction

Introduction

Recent interest in finding renewable sources of gaseous and liquid fuels has been particularly strong in connection with agriculture. Because agriculture is both a producer and consumer of energy, agriculturally derived fuels offer the attractive option of tightly integrating the production, transformation, and consumption stages of the fuel cycle. Thus, farmers could benefit by taking on an additional production activity as well as by reducing their dependence on a purchased input whose supply and price outlook is particularly problematic. Moreover, even if the fuel is intended for use in the overall economy rather than on the farm, agriculture is a logical place to seek alternatives to fossil fuel. It occupies a large fraction of the nation's land and has an exceptionally well developed infrastructure: a marketing system, storage and processing facilities, and a transportation network. If we are concerned with crops that are already grown for non-energy purposes (food, feed, and fiber), an additional advantage is that producers have a substantial body of knowledge and experience. These features are shared by only a few of the non-agricultural biomass materials now under consideration.

Thus it is not surprising that in the past few years there has been a sharply increased activity relating to several agriculturally based fuel sources, including carbohydrate crops for ethanol, livestock manures for biogas (methane), crop residues for producer gas, boiler fuel, or ethanol, and vegetable oils as a diesel fuel. All of these have a long history in some part of the world, but in the United States interest in vegetable oil fuels is a particularly recent development. This interest has focused both on unfamiliar crops and on crops that are very well established, such as soybean.

177

Virtually all research and development projects involving familiar oilseeds have emphasized technical considerations, such as the best chemical and physical form of the oil, its effects on engine wear, and end-use efficiency and atmospheric emissions. Little attention has been given to economic factors beyond a simple comparison of the current price of the vegetable oil to that of diesel fuel. Typically, a particular oilseed is examined by itself, with no analysis of how the various types compare. This is satisfactory if only one type is produced in a particular area; the potential national market is so great that no type can preempt another type grown elsewhere. However, crop to crop comparisons are very relevant when two types compete for the same land. Such competition may occur in the North Central (Corn Belt) region, which is now a major soybean producing region, but which may also see expanded sunflower production. Sunflower was a crop of negligible significance until the middle of the last decade, but in four years became a major crop. The prospect of using sunflower oil as a fuel may stimulate even further expansion into areas adjacent to the original producing areas (Northern Plains and Texas), including the western Corn Belt.

Another inadequacy of simply using current price data is that use of vegetable oils as a fuel will itself change the supply, demand, and price situation for oil and oilseeds. Moreover, in the case of on-farm processing and use, current prices are only partially relevant, since they refer to the present system in which oilseeds are processed in centralized facilities. Finally, on-farm processing has significant implications for other aspects of farm management, since it provides not only a fuel but also a high-protein livestock feed, whereas at present farmers purchase these feeds from the oilseed mill industry.

This paper examines the economic attractiveness of various oilseeds in a more systematic way than simply by using current prices. The major oilseeds now produced in the United States are examined from several viewpoints: 1) Ratio of value of the oil to the non-oil (feed) value; 2) Implications of this ratio for the price and supply response to an increased demand for the oil for fuel; 3) Implications for the comparative suitability of soybean and sunflower for on-farm processing. Because so many criteria enter into a comparison of various oilseeds no one type emerges as clearly preferable. It will be shown that by considering the entire seed, including both oil and non-oil components, the attractiveness of various oilseeds changes substantially compared to an analysis based purely on current prices.

Current Economic Conditions For
Vegetable Oils

Production

This paper deals with four major oilseeds plus corn, which together represent close to the entire U.S. production of edible vegetable oils. Corn is included for illustrative purposes, even though it is not an oilseed and its oil is not being considered as a diesel substitute. Indeed, the very fact that it is not under consideration illustrates the importance of examining the entire crop, not just the oil component. From just the production and price data of Tables 1 through 3, one could conclude that corn oil is at least as plausible a diesel fuel substitute as other oils under consideration. The problem is that corn is primarily a source of products other than oil, and overlooking this aspect would lead to a serious misinterpretation of the price and production data. This same misinterpretation could occur in analyses of oilseeds.

Table 1 shows that soybean oil accounts for about four-fifths of total U.S. edible vegetable oil production, a proportion that has held almost constant during the considerable increase in total output in the past decade. The closest competitor, cottonseed, accounts for less than one-tenth of the total. Its share has been continuously declining because the output of cottonseed oil has been roughly constant. Sunflower oil, while still contributing less than 5% of the total, is particularly interesting because of its sudden rise to prominence. In contrast to more established oilseeds, sunflower has only begun to achieve its eventual place in the overall picture.

Table 1 refers only to the oil actually extracted in the U.S. from each crop; the part of the crop used in other ways potentially represents a source of additional oil. Table 2 shows that substantial portions of the soybean and sunflower crop are exported and could help meet an increased domestic demand. Cottonseed, in contrast, is almost entirely consumed domestically. Peanut is raised primarily for human consumption, a high value market with which non-food uses would have difficulty competing. Corn is both exported and used domestically for feed. The problem with shifting either of these uses to oil extraction is that only a small fraction of the value of the corn is found in the oil, so that the process is limited by the markets for the co-products of corn wet milling (corn starch, gluten feed, and gluten meal).

Table 1.   Domestic Production of Edible Vegetable Oils
            (million metric tons)

Year Be-ginning Fall	Soybean	Sun-flower	Cotton-seed	Peanut	Corn	TOTAL[a]
1970	3.75	---	.56	.12	.22	4.65
1971	3.58	---	.59	.12	.23	4.52
1972	3.40	---	.71	.12	.24	4.47
1973	4.08	---	.70	.09	.24	5.11
1974	3.35	.04	.61	.11	.21	4.32
1975	4.37	.05	.42	.22	.29	5.35
1976	3.89	.02	.54	.14	.30	4.89
1977	4.67	.09	.66	.07	.33	5.82
1978	5.14	.12	.58	.07	.33	6.24
1979	5.49	.22	.65	.09	.36	6.81
1980	5.11	.30	.54	.07	.40	6.42
1981	4.97	.14	.71	.08	.40	6.30

[a]Does not include linseed oil (an industrial oil) or safflower oil, which accounts for less than 1% of total U.S. vegetable oil production.

Source: Fats and Oils, Outlook and Situation, various issues, 1980-1982; Foreign Agriculture Circular, Oilseeds and Products, FOP 22-81 (December 1981). U.S. Department of Agriculture.

Table 2. Uses of Oilseeds as % of Total Utilization[a]

| | Processed for Oil | | | Other Domestic Uses | Exports |
	Domestic Oil Use	Export-ed Oil	Total		
Soybean	42	9	51	4	45
Sunflower	8	10	18	8	74
Cottonseed	36	40	76	23	1
Peanut	13	3	16	68	16
Corn (est.)	6	2	8	63	29

[a] Based on 1981 crop. Imports of all crops in table are negligible. Computed from data in Agricultural Statistics, 1980; Fats and Oils, Outlook and Situation, FO5-309 (October 1982). U.S. Department of Agriculture.

Prices

Wholesale prices of the major vegetable oils (tank-car quantities, at the oilseed mill) are shown in Table 3. For comparison to diesel fuel, 1 kg of vegetable oil can be taken as about equivalent of 1 liter of diesel fuel (1), a fortunate coincidence, since vegetable oil data typically are given gravimetrically while diesel fuel is reported volumetrically. By analogy with diesel fuel, we may assume that the mill to farm mark-up would come to a few cents per liter. (No farm-level market currently exists.) If so, the cheapest vegetable oil (soybean) would have been almost twice as expensive as farm-delivered diesel fuel in 1981, which typically was $0.31/liter.

A prominent feature of vegetable oil prices is their sharp year-to-year fluctuations. In contrast to some other renewable energy systems whose costs (in constant dollars) are likely to change smoothly over time, vegetable oil prices are too volatile to predict with confidence. Thus one cannot say, as is commonly done with other fossil fuel alternatives, that expected fossil fuel price increases will eventually make them economically competitive, even if they are not at present.

Table 3.   Prices of Crude Vegetable Oils[a]

Calendar Year	Wholesale Price ($/kg)					Index of All Oils
	Soybean	Sun-flower	Cotton-seed	Peanut	Corn	
1970	.27	---	.30	.42	.36	100
1971	.28	---	.34	.45	.44	106
1972	.23	---	.25	.45	.36	90
1973	.44	---	.43	.57	.50	145
1974	.79	---	.84	1.09	.90	249
1975	.56	---	.60	.86	.71	185
1976	.41	---	.51	.69	.57	136
1977	.52	---	.54	.66	.68	161
1978	.57	.59[a]	.60	.93	.79	173
1979	.61	.73[a]	.70	.82	.71	189
1980	.52	.56	.55	.68	.58	141
1981	.47	.58	.53	.83	.52	123
Coeff. of Corr. (r) with Index	.96[b]	---	.95	.88	.95	---

[a] 12 months beginning previous October.

[b] High correlation of soybean with index of vegetable oil prices is an automatic consequence of the fact that the index is comprised mainly of the price of soybean oil.

Sources: Agricultural Statistics, 1974, 1978, 1980; Fats and Oils, Outlook and Situation, various issues, 1980-1982; Foreign Agriculture Circular, Oilseeds and Products, FOP 3-82 (February 1982).   U.S. Dept. of Agric.

A second striking feature is the consistent pattern of
relative prices among the various kinds of oil even though
their prices as a group vary sharply (as shown by the correlation of individual prices with the index of vegetable oil
prices). Soybean oil is consistently the cheapest, with
peanut and corn oil almost always the first and second most
expensive kinds respectively. This price pattern reflects
the fact that the various oils are largely interchangeable,
but with certain differences in their use that create a premium for certain kinds. This means that any analysis of the
future economics of vegetable oil as a diesel fuel should not
deal with an individual type in isolation, but rather consider the entire vegetable oil market. For example, since
sunflower is a crop whose potential has just begun to be
tapped, it is conceivable that it can be produced more
cheaply than at present. However, the price of sunflower oil
will always be linked to that of soybean oil. This point
will be explored further in the discussion of the possible
production response of sunflower to higher oil demand.

Value of Oil and Non-Oil Products

Extraction of oil from these crops also produces a
high-protein livestock feed in the case of the four oilseeds,
and a variety of products in the case of corn. Moreover, the
cottonseed itself is a by-product since the lint has a higher
value than the seed. Table 4 shows the relative importance
of the oil in each case, expressed both as a fraction of the
farm price of the crop and of the total value of all products
(at wholesale prices). The oil has the least relative importance for corn and cottonseed, for reasons already discussed,
and the greatest importance for peanut and sunflower. If
cottonseed is considered without the lint, it would be similar to sunflower and peanut. Soybean is an intermediate
case: even though it is considered an oilseed, the oil
actually accounts for less than half of the total value.

The two columns in Table 4 are closely related, but
show somewhat different patterns because of differences in
the farm to processing plant price spread for the seed and
also in the value added in processing the different crops.
Both of these are included in the denominator of the right-
hand column. (The numerators of both columns are the same.)
Thus for corn, which undergoes complicated processing to
yield high value co-products, the value of the oil is only
14% of the total value of products, compared to 28% of the
farm value of the grain. On the other hand, processing of
soybean is a highly efficient operation that adds only a
small amount to the value of products compared to that of
the crop, typically about 10%.

Table 4.   Value of Crude Vegetable Oil Compared to Value of
Crop and Non-Oil Products [a]

	Ratio of Value of Oil to Farm Price of Crops	Ratio of Value of Oil to Value of All Products
Soybean	.45	.40
Sunflower	1.22	.77
Cottonseed	.12	.11
Peanut	1.06	.76
Corn	.28	.14
Ethanol from Corn[b]	1.33	.77

[a] Based on 1978 crop. Soybean, sunflower, and peanut includes oil and meal; cotton includes lint, oil, meal, linters, and hulls; corn includes oil, gluten meal, gluten feed, and starch. Products valued at average wholesale price, Fall 1978 to Fall 1979. Sources: Agricultural Statistics, 1980; Fats and Oils, Situation and Outlook, various issues, 1980-1981; Feed Outlook and Situation, various issues, 1980-1981. U.S. Department of Agriculture.

[b] Estimate only because of varying price of ethanol. Based on yield of 370 liters of EtOH and 300 kg distillers dried grains and solubles per metric ton of corn. Corn at $100/metric ton; DDGS at $133/metric ton; EtOH at $0.36/liter. Value of "oil" is that of EtOH.

The left-hand column is relevant to the question of how much an increased demand for oil will in turn stimulate an increase in production of the crop. When the ratio is low, as with corn or cottonseed, there is little effect, i.e., the proportional change in the value of the crop is much smaller than the change in the price of oil. With peanut and sunflower, in contrast, the feedback is very strong, since these crops truly are oilseeds. Soybean is an intermediate case.

The right-hand column relates to the question of how much the markets for non-oil products from a given crop are affected by increased processing in response to a higher demand for oil. The ratio of the physical quantities of oil and non-oil components is virtually fixed. Consequently, for each additional dollar's worth of oil produced, other products worth anywhere from $0.31 (peanut) to $8.45 (cotton) must also be sold. Depending on the elasticity of demand for the various non-oil products, this will depress their prices. The demand and price of the whole seed are determined by the demand and prices of all products, along with processing costs. Consequently, there will be little or no price stimulus to increase production in the case of cotton or corn. In contrast, this problem is much less important with sunflower and peanut. The order of the crops' relative attractiveness in this regard is the same regardless of which of the two measures is used; however, because of the low processing costs with soybean, the gap separating soybean from sunflower and peanut is not so great when the oil is compared to the value of all products as when it is compared to the farm price of the crop.

For comparison, Table 4 shows analogous ratios for ethanol production from corn grain, with the value of the ethanol fuel used instead of that of oil, and with distillers dried grains and solubles (DDGS) the by-product. This case is comparable to the two most favorable oilseeds (sunflower and peanut). Yet marketing of distillation by-products is considered an important factor affecting the economic attractiveness of ethanol from grain (2-4). Consequently, Table 4 suggests that the analogous problem will be quite serious with soybean.

This analogy is even more plausible in that DDGS is a high-protein livestock feed that would be sold largely to the same markets as soybean meal, so that soybean oil and ethanol fuel both face the same market limitations. Moreover, it suggests that simultaneous adoption of both technologies will be particularly difficult, since an expansion of ethanol production is expected to reduce the area planted to soybean (2, 3).

Potential Expansion of Vegetable Oil Production for Fuel Use

Compared to its potential use as a diesel fuel, national production of vegetable oil is quite small. The total domestic annual production of 6 to 7 million metric tons (Table 1) is equivalent to slightly under 2 billion gallons of diesel fuel. Including exported oilseeds brings the total to about 3 billion gallons, or slightly less than the total quantity of diesel fuel used in agricultural production, estimated at 3.3 billion gallons in 1978 (5). Since at most a small fraction

of existing oil supply might be diverted to fuel use, clearly
this technology will not make a significant contribution to
the national energy situation unless total production of vegetable oils can be increased very sharply. This section
explores this possibility under the assumption that production
and consumption are governed by interactions among price,
supply, and demand. Alternatively, the determining factor
might be subsidies, tax incentives, or other forms of government intervention, as was done to stimulate alcohol fuel production starting in the late 1970's.

Soybean

In the short run, only soybean offers any prospect of
supplying a significant quantity of oil for fuel use. For a
very simple analysis of what in reality is a highly complex
market, we can assume that the soybean supply is governed by
a price elasticity e_s and that demand for soybean meal and oil
are governed by price elasticities $-e_m$ and $-e_o$ respectively.
If the processing margin (total value of products less the
price of the soybeans) is assumed to stay fixed, then the
prices of the soybean products and of soybean are not independent. It is convenient to express the price of meal and oil
as fractions V_m and V_o of the farm price of soybean. Typically $V_m = 0.7$ and $V_o = 0.4$ (Table 4), with the products
together worth 1.1 times the farm price.

Let us assume that before any soybean oil is used as a fuel,
the prices of oil, meal, and soybeans are in equilibrium.
Using oil as a fuel then increases the demand for oil, in
turn increasing its price and correspondingly the price of
soybeans. The resulting increased production of both meal
and oil lowers the price of the former, partially offsetting
the rise in the price of the latter, leading to a new equilibrium.

Because the fuel market could absorb any conceivable
level of soybean oil production, some assumption is needed to
define the point at which fuel use stops. For example,
suppose that use as fuel starts when the price of diesel fuel
equals that of an equivalent amount of soybean oil, and that
fuel use grows until soybean oil is 10% more expensive. (For
example, users may perceive some non-price benefits in using
soybean oil.) When the new equilibrium is reached, S_f, the
fraction of the initial soybean oil supply now used as fuel,
is given by

$$S_f = \left[(P'/P) - 1\right] \times e_o \times (1/e_s + V_m/e_m + V_o/e_o) \; / \; (1/e_s + V_m/e_m)$$

where P and P' are the old and new prices of soybean oil.

Unfortunately, there are great uncertainties in the elasticities of supply and demand of soybean and soybean products respectively. For illustrative purposes, we may use intermediate values typical of the range reported in the literature (6,7): e_s = 0.5 (moderately inelastic supply); e_o = 0.1 (highly inelastic demand for oil); e_m = 0.5 (moderately inelastic demand for meal). A very important simplification is that the entire soybean market (domestic plus export) is considered together as a single market.

Under these assumptions, a 10% increase in the price of oil occurs with soybean production increasing by 1.2% in response to a 2.4% increase in the farm price of soybean. Soybean oil consumption for non-fuel uses decreases by 1% (because of a non-fuel elasticity of -0.1), so that 2.2% of the original supply is available for fuel use.

Thus the adoption of this technology is highly self-limiting; once it starts to be competitive, its use feeds back on the initial price situation to make it uncompetitive rather quickly. Even though the details of the above calculation are highly uncertain, nevertheless no reasonable combination of supply and demand elasticities will result in a large quantity of soybean oil being available for any new use such as fuel without sharply increasing its price. This is quite different from many other renewable energy systems, such as solar collectors, for example, where supply elasticity is not a problem. Once a collector is economically competitive, more and more can be made while still remaining competitive. In fact, to the extent that there are economies of scale, initial adoption of the technology provides a positive feedback and stimulates further adoption because production costs are lowered. In contrast, for a well-established crop like soybean, higher prices are needed to increase production, all else remaining the same.

The reason for the poor supply response to a higher demand is found not only in the limited supply elasticity (which would apply to all well-established agricultural crops), but also in the additional meal production that is a necessary consequence of expanded oil production. In the example above, the 1.2% increase in soybean meal production causes a 2.4% decrease in the price of meal (because of the assumed elasticity of -0.5). Because the meal is worth 0.7 times the farm price of soybeans, this lowers the derived value of the soybean by 1.7%, substantially offsetting the 10% increase in the price of the oil, which contributed a 4% increase in soybean value. Simply stated, when demand for some product increases, producing more of a crop that primarily yields some other product is not an effective response, especially when the market for the main product is elastic.

Sunflower

In contrast to soybean, sunflower's main value is found in the oil, with the meal constituting a by-product. This suggests that if demand for oil is increased because of a new use, i.e., fuel, the resulting market shifts would favor sunflower over soybean. For sunflower, V_o is about 1.2 compared to 0.4 for soybean, while V_m is only 0.3 compared to 0.7 for soybean. If sunflower were governed by the same relationships as soybean, then 5.6% of the oil supply would be usable as fuel under the assumed limit of a 10% increase in the price of oil. Sunflower production would increase by 4.6% while non-fuel consumption would again decrease by 1.0% (a consequence of the assumed elasticity of demand for oil, independent of the particular type). The increased meal production depresses the value of the seed by 2.8% whereas the higher price of oil increases it by 12%. Thus the effective price is increased by 9.2%, leading to a production increase of 4.6% (under the assumed supply elasticity of 0.5, as with soybean).

Of course, it is unrealistic to apply to sunflower the same analysis as was applied to soybean, since sunflower is a minor part of the overall oilseed picture at present, whereas soybean dominates it. Moreover, 5.6% of current sunflower oil production is much less oil than 2.2% of soybean oil production. (These are the respective proportions available for fuel use under the same assumed oil price increase of 10%.)

Eventually, however, sunflower may have a major place in the total vegetable oil supply. Its production at present is limited by non-economic factors; it is an unfamiliar crop and has not yet been tried in all the areas where it might be expected to be competitive. The small current sunflower oil supply largely goes to specialized uses where it commands a premium, as shown in Table 3. If the supply is expanded severalfold, sunflower oil would serve as a general-purpose vegetable oil, a role now taken by soybean oil, and the prices of the two would be even more closely linked than at present (Table 3).

If this happens, then a demand for any vegetable oil as fuel would tend to favor sunflower over soybean, regardless of which type is actually used as fuel. At present, there is only one major vegetable oil, so that two products -- oil and meal -- are supplied in essentially fixed proportions, except

to the extent that we also export individual products along
with whole soybeans. With two oilseeds, a new degree of
freedom is introduced, and a shift in relative demand can be
accomodated not only by adjusting the total output -- the only
degree of freedom now available -- but also by shifts within
the oil sector. These shifts need not take place on the same
land, although this would be one possibility if sunflower
production expands into soybean areas of the Western Corn
Belt. Through less direct market interactions a reduction in
soybean production in the Corn Belt could occur along with an
increase in sunflower production in the Plains. The mechanism for this could be that soybean land is reduced in the Corn
Belt in favor of corn, a feed grain, while sunflower is expanded in the Northern Plains at the expense of barley, or in the
Southern Plains at the expense of sorghum, both of which are
feed grains that serve the same market as corn. In analyses
of this type, it is important to remember that prices of the
major crops like oilseeds and feedgrains interact with each
other through the national market, so that they need not be
competitors for the same land to affect each other's supply
and price. Thus paradoxically, use of soybean oil as a
diesel fuel might reduce soybean production and increase sunflower production, since sunflower supplies more oil, the
product that now is in relatively greater demand. Of course,
none of this will apply unless sunflower becomes a major oilseed, with a fully developed national market in which it is
fully interchangeable with soybean oil.

On-Farm Processing

The discussion thus far has been based on oilseed processing in centralized facilities, which is the current situation. An alternative that is also under major consideration
is for the farmer to process the crop on-farm. The advantages
usually put forth for this approach are that it frees the
farmer from dependence on an external source of fuel, thereby
providing protection against a fuel shortage, and that it
eliminates the marketing costs for the oilseed and the wholesale to farm price markup for the oil. The penalty is that
on-farm facilities are not as efficient as centralized ones
because they cannot use solvent extraction, and instead must
use mechanical extraction. Consequently, the oil extraction
rate will typically be 15% for soybean and 32% to 36% for sunflower, in contrast to 18% and 40% respectively for centralized facilities (8). The unextracted oil remains in the meal,
which is therefore more susceptible to spoilage.

Table 5 shows the price advantage in using one's own
soybeans and in consuming the products, compared to a centralized facility. Three transactions are saved: selling the soy-

Table 5.   Comparison of Extraction of Crude Soybean Oil at
          On-farm and Centralized Facilities[a]

	On-farm	Central- ized	On-farm Advantage (Disadvantage)
Price cf soybean meal ($/metric ton)	289.46	210.10	79.36
Product yields (weight per weight of soy- beans)			
Oil	0.15	0.183	(-0.033)
Meal	(b)	0.794	(b)
Production expenses ($ per metric ton of soybeans)			
Soybeans	247.59	254.23	6.63

[a]Based on average prices and product yields, 1976-1980.
Wholesale prices are for Illinois; farm price of soybeans
is national average. Sources: prices from Foreign
Agriculture Circular FOP 22-81, Oilseeds and Products
(December, 1981); product yields and farm price of meal
from Fats and Oils Situation, various issues, 1978-81.
U.S. Department of Agriculture.

[b]On-farm meal yield is slightly higher because of higher
oil content. The value of meal per ton of soybeans is
taken as equal in the two cases. This ignores the digesti-

Table 5, continued.

	On-farm	Central-ized	On-farm Advantage (Disadvantage)
Processing [(c)]	C_p	11.13	$(11.13-C_p)$
Credit for value of meal	-229.83	-166.58	63.25
Cost of oil at facility	$17.76+C_p$	98.78	$81.02-C_p$
Cost of oil ($/kg)			
At facility	$0.12+C_p'$	0.54	$0.42-C_p'$
Wholesale to farm markup	0	~ 0.03	~ 0.03
At farm	$0.12+C_p'$	~ 0.57	$\sim 0.45-C_p'$

ble energy value of the additional oil, and also the greater susceptibility to spoilage of farm-produced meal.

[(c)] C_p is the on-farm processing cost per ton of soybeans; C_p' is the cost per kg of oil produced ($C_p/150$ for the assumed yield). "Cost" of oil in the on-farm case is imputed; in the centralized case it is the actual wholesale oil price, whereas processing cost is imputed from value of products less price paid for soybeans by the facility. See text for discussion of the different principles underlying the two calculations.

beans, purchasing the oil, and purchasing the meal (assuming
the farm uses meal). Of these three, the last is by far the
most important, since the oilseed mill to farm markup for
meal is far greater than the farm to mill markup for soybeans.
The mill to farm markup for oil is not known, since farmers
do not currently purchase soybean oil, but by analogy with
diesel fuel it is probably only a few cents per kg.

Offsetting these advantages are the lower extraction rate
as well as the presumably higher processing costs (C_p in
Table 5). If on-farm processing costs are less than six times
as great as those of centralized facilities--a limit that
seems readily attainable--then farm-produced oil is cheaper.
(C_p = \$68/metric ton, or C_p' = \$0.45/kg, leading to a farm
price of \$0.57/kg.) For on-farm soybean oil to be competitive
with diesel fuel at \$0.31/liter, extraction costs plus the
cost of upgrading from the crude oil of Table 5 must be less
than \$35/metric ton of soybeans, or \$0.23/kg of oil. There
is a disagreement among researchers on the necessary degree
of upgrading, so that a cost cannot be assigned unambiguously.
In fact, it is not clear that the necessary processing could
be done satisfactorily on-farm; this issue is currently
receiving considerable research attention (9).

Although the two cases are presented in parallel in the
table, in fact the underlying logic is quite distinct. With
centralized facilities, soybean oil has a market price, as
does the meal. The price of soybeans is derived from these
two prices, reflecting the yields of the respective products
along with processing costs. With on-farm processing, however,
the soybeans have a value determined by the market, as does
the meal. The oil's cost is derived from these; i.e., it is
the residual after credit for the meal is deducted from what
the farmer would have gotten for the soybeans, plus processing
costs. This means that the cost of the oil is even more
volatile than the price of commercially purchased oil, since
it is highly sensitive to the market for meal. In contrast,
the commercial market for oil is independent, except to the
extent that a changed demand for meal affects the price and
hence the supply of soybeans, and consequently of soybean oil
as well.

Although in the long run the prices of meal and of soybeans are strongly coupled, in the short run price fluctuations for the two are sometimes out of synchrony. In some
years, the cost of farm-produced soybean oil when computed
according to the method of Table 5 would have varied from
substantially higher than that of commercial oil to negative.
The latter occurs when the meal is worth more to the farmer

than the soybeans from which it was produced, plus processing
costs (1).

Sunflower

A similar calculation for sunflower is shown in Table 6,
although certain changes in the calculations were necessary
because of differences in the available data on sunflower and
soybean. On-farm processing costs additional to the cost of
sunflower and the credit for the meal are those of a plant
that processes 1.5 metric tons/day. This was the most economical of three sizes analyzed for production of 16,000 kg of
oil per year, the diesel requirements of an average North
Dakota farm (10). A smaller press would be used for the
Minnesota farms considered later, so that production costs
would be higher and the extraction rate lower (32% instead of
36%, whereas centralized plants achieve 40%). Because the
resulting meal has 11% oil content (18% with the smaller
plant), its value may be considerably lower than that of
commercial sunflower meal, even after allowance for the lower
protein concentration. The table reflects the uncertainty in
this effect; with soybean, this effect was ignored, and the
main uncertainty was in the processing cost.

Unlike with soybean, on-farm processing of sunflower is
not more economical than centralized processing. The reason
is that sunflower contains less meal, and the plant to farm
markup in meal was the main source of on-farm soybean processing's advantage. Yet this same characteristic will permit
sunflower to meet a larger fraction of a given farm's fuel
needs without producing more meal than can be consumed on-farm.
As shall be seen, this could even mean complete replacement of
diesel fuel if technical considerations will permit this (11).
The value of being completely independent of external supplies
of diesel fuel could then be credited to on-farm sunflower
processing. In contrast, the fuel security value of soybean
is negligible if only partial replacement is possible because
of excessive meal production. Nevertheless, on-farm production of all fuels, including soybean oil, is sometimes spoken
of as a goal well worth striving for regardless of the degree
of self-sufficiency that can be achieved.

On-Farm Use of Vegetable Oil and Meal

The main economic advantage of on-farm processing of
soybean lies in the saving in the cost of purchased soybean
meal. For this advantage to be captured, the farm must be
able to use all the meal produced along with oil for fuel.
Whether a particular farm can do so depends on the relative

Table 6. Comparison of Extraction of Crude Sunflower Oil at
On-farm and Centralized Facilities[a]

	On-farm	Central-ized	On-farm Advantage (Disadvantage)
Price of sunflower meal ($/metric ton)	153[b]	110	43
Product yields (weight per weight of sunflower)			
Oil	0.36	0.40	(-0.04)
Meal	(c)	0.60	(c)
Production expenses ($ per metric ton of sunflower)			
Sunflower seed	227	233[d]	6

[a] Based on average prices and product yields (centralized
facilities), 1978-80. Wholesale prices are for Minneapo-
lis; farm price of sunflower is average for major producing states. Prices from Fats and Oils Outlook and
Situation, FOS-306 (February 1982). U.S. Department of
Agriculture.

[b] Assumed price of sunflower meal of same quality as commercial meal, if plant to farm markup is same proportion as
with soybean meal(39%).

[c] On-farm meal yield is slightly higher because of higher
oil content. The fractional reduction in the value of the
meal (compared to commercial meal with the same total
quantity of protein) is shown below as the unknown D_m. In
this representation, D_m would also reflect any inaccura-

Table 6, continued.

	On-farm	Central-ized	On-farm Advantage (Disadvantage)
Processing	$130^{(e)}$	83	(-47)
Credit for value of meal	$92 \times (1-D_m)$	66	$26-92 \times D_m$
Cost of oil at facility	$265+92 \times D_m$	250	$(-15-92 \times D_m)$
Cost of oil ($/kg)			
At facility	$0.74+.26 \times D_m$	0.63	$(-0.11-.26 \times D_m)$
Wholesale to farm markup	0	~ 0.03	~ 0.03
At farm	$0.74+.26 \times D_m$	0.66	$(-0.08-.26 \times D_m)$

cies in the assumption given in (b) regarding the farm price of commercial sunflower meal.

(d) Prices paid by sunflower processors are not available. This price was calculated from the prices received by farmers, assuming the same farm to plant markup as with soybean. The $6/metric ton markup was deducted from the calculated value of products less farm price of sunflower to give processing costs of $83/metric ton. The value of $250 of oil per metric ton of sunflower processed is actually an input in this calculation, not a derived value. (See discussion of Table 5).

(e) Derived from data given in Ref. 10, for 1.5 metric ton/day plant producing 16,000 liters of sunflower oil per year. Includes estimated cost of utilities, depreciation and repairs on building and equipment, insurance, and labor.

importance of crop production, the main use of diesel fuel,
and of animal production, particularly of hogs, the main consumer of soybean meal. Poultry are also an important consumer
of soybean meal, but since most poultry production is concen-
trated in specialized facilities separate from crop production,
it is not relevant here.

With sunflower, the advantage of an on-farm meal supply
is not as great, since less meal is produced. Nevertheless,
the farm will still have to use the meal for the process to
be economically attractive, because established markets for
meal in this quantity do not exist. In some locations, a
multi-farm arrangement could solve this problem; however,
this paper considers only the single-farm case.

Table 7 shows how this applies to two representative
farms in Minnesota, the only state that currently is a major
producer of both soybean and sunflower. The table shows the
average number of hogs and the average diesel fuel use for
livestock farms and cash grain farms that use diesel fuel and
that have hogs. (The other farm characteristics are for all
farms of the respective farm types, not just those that use
diesel or that have any hogs.)

The table also shows how much sunflower and soybean
processing could occur on each type of farm if all the resulting oil is used to replace diesel fuel. For example, at
assumed on-farm oil extraction rates of 15% and 32% respectively for soybean and sunflower, either 53 metric tons of
soybean or 25 metric tons of sunflower would yield 7.9 metric
tons of oil, the equivalent of the 7,900 liters of diesel
fuel actually used on the average livestock farm. Similarly,
25 metric tons of soybean or 52 metric tons of sunflower
would yield 20 metric tons of soybean meal or 31 metric tons
of sunflower meal, which are equivalent in protein. For
national average feeding rates for hogs, either of these
would supply the soybean meal used by the average inventory
of 202 head on Minnesota livestock farms with hogs. In
reality, this much sunflower meal could not be used, because
of its high oil content and low lysine content (8). However,
at least in the case of the livestock farm, the farm's ability to use oil is what will limit the quantity of sunflower
processed, and only partial replacement of soybean meal by
sunflower meal will be necessary. For both crops and both
kinds of farms, whether the ability to use the meal or the
oil is the limiting factor, the quantity that can be processed
is well below the average amounts raised on farms that already
are raising the particular crop.

An important difference between sunflower and soybean is

the order in which the two limits are in. Even for a livestock farm, which by definition has a high livestock to crop
ratio, full replacement of purchased soybean meal by farm-
produced meal results in less than half of diesel fuel use
being replaced. With cash grain farms that also have hogs,
the fraction obviously is even smaller.

With sunflower, the proportions are almost exactly
reversed: if all diesel fuel used on livestock farms is
replaced by farm-produced sunflower oil, the resulting supply
of meal is the protein equivalent of just under half of
current soybean meal consumption by hogs. Even on cash grain
farms, with a much higher crop/livestock ratio, the two products are produced in just about the right proportions to be
used on-farm. (Recall, however, that complete replacement
of soybean by sunflower meal with hogs is not desirable, and
complete replacement of diesel fuel by vegetable oil may
never be made technically feasible.)

This means that with soybeans, farms of the types shown
in the table cannot achieve complete diesel fuel self-sufficiency while also using all of the co-produced meal. Yet the
latter is essential if the full economic benefit of on-farm
processing is to be obtained. On the other hand, if the farm
can replace only a portion of its external fuel supply, the
limited fuel supply security that results could be achieved
equally well simply by increasing the farm's fuel storage
facilities by a corresponding amount. This point is often
glossed over by some advocates of on-farm fuel production,
whether ethanol or vegetable oil, who sometimes portray even
a fractional reduction in external fuel dependence as so
valuable in an age of threatened supplies that it should be
an objective regardless of the cost.

In contrast to soybean, with sunflower it indeed may be
possible to achieve complete diesel fuel self-sufficiency
while simultaneously making full on-farm use of the meal,
assuming the necessary technical advances in vegetable oil
are achieved, and assuming the sunflower meal can be substituted for soybean meal at such a high rate in hog diets.
Also, the result will depend on how a particular farm's
livestock/crop ratio compares to those of the two important
types shown in the table. But in any case, sunflower will
always have the same relative attractiveness compared to soybeans as was shown in the table, as a consequence of having
an oil to protein ratio about 4.5 times that of soybeans.
This is the same characteristic that makes sunflower more
attractive for expanding oil production: if you are interested in obtaining the oil, it is preferable to grow a crop
that has more oil and less of some other product that still

Table 7.   Potential Use of Soybean and Sunflower for Fuel
and Feed on Two Kinds of Minnesota Farms, 1978[a]

	Livestock Farms	Cash Grain Farms
Farm Characteristics[b]		
Size (hectares)	88	141
Crops (hectares)		
Soybean	35 (26%)	44 (63%)
Wheat	19 (11%)	49 (40%)
Corn	39 (57%)	45 (73%)
Oats	12 (44%)	17 (39%)
Sunflower	51 (0.6%)	85 (6%)
Hogs	202 (52%)	115 (14%)
Diesel fuel use (1000 liters)	7.9 (46%)	10.3 (71%)

[a] Farm characteristics from Census of Agriculture, 1978,
Minnesota. Data are averages for farm types 021 (livestock) and 011 (cash grain).

[b] Data are for farms with indicated activity; figures in
parentheses are percent of farms with that activity.

[c] Meal useable by hogs for national average consumption of
soybean meal per head of hog; sunflower meal assumed use-

Table 7, continued.

	Livestock Farms	Cash Grain Farms
Maximum On-Farm Use of Oilseeds (Metric tons/farm)		
Soybean		
Limited by on-farm use of meal [c]	25	14
Limited by on-farm use of oil [d]	53	69
Actual production [e]	79	96
Sunflower		
Limited by on-farm use of meal [c]	52	28
Limited by on-farm use of oil [d]	25	32
Actual production [e]	79	135

able on equivalent protein basis (see text for qualifications).

[d] Oil useable assuming complete replacement of 1978 average diesel fuel use (per farm using any diesel) on equivalent energy basis.

[e] Based on yields for each type of farm in 1978 and average area in indicated crop per farm raising the crop.

must be used (or marketed) to make the economics of the overall process competitive.

Summary and Conclusions

The economic attractiveness of various vegetable oils as
fuels must be determined on the basis of many criteria. No
single type among those now produced in the United States
emerges as clearly preferred when all criteria are applied
simultaneously. Currently, soybean oil is by far the leading
vegetable oil in terms of domestic production, and its potential importance is further increased by the fact that a large
fraction of the crop is exported in addition to the amount
now used for domestic production of oil. Of the less important oils, only sunflower might challenge the position of
soybean, since it is a new crop that has great potential for
being grown outside its current production areas. The other
oilseeds (peanut, cottonseed) have not shown any significant
increase in recent years.

Soybean is also the clear favorite with regard to current
prices, although here too a direct comparison with sunflower
overlooks the fact that sunflower is still a relatively
specialized oilseed, with the oil going to premium markets.
If sunflower production expands sharply, the price differential would be eliminated, since the larger market would no
longer regard sunflower as deserving a premium. In fact,
the original statement should be turned around: sunflower
production will expand only if production costs can be kept
sufficiently low to permit production to be profitable when
the oil sells at about the same price as soybean oil.

An important factor frequently overlooked in economic
analyses of vegetable oil fuels is the joint production of
one or more co-products along with the oil. Unless the markets for these products are strong, the price-depressing
effect of increased production will offset much of the price-
increasing effect of higher demand for oil, which in turn
translates into only a small increase in the demand for the
oilseed. Thus one cannot simply take the current market
prices of a particular oilseed and its products and assume
they will remain constant as production increases; the feedback effects of supply and demand elasticities limit the
potential supply of vegetable oil that can be expected at a
given price, e.g., a price competitive with diesel fuel. The
higher the value of the oil as a fraction of the value of the
crop (or of the total value of all products), the greater the
supply response to an increased demand for oil. Sunflower and
peanut oil are clear favorites in this respect, although the

static supply and higher price of peanut oil are great offsetting disadvantages. Cottonseed oil can be virtually ruled
out because the economics of cotton production are determined
primarily by the market for lint. Soybean is an intermediate
case, with a meal to oil ratio higher than that of sunflower
or peanut.

The meal to oil ratio also determines the potential for
on-farm processing of oilseeds. With soybean, the main economic advantage of this approach is that it saves the farmer
the need to buy soybean meal, for which the markup over the
price at the processing plant is very significant. Consequently, the farmer must be able to use all the meal that is
produced. Only farms with very large livestock inventories
in relation to crop production will be able to do this if
they also produce all the diesel fuel substitute they can use.
More typical farms will be able to replace only a fraction of
their diesel fuel by soybean oil if they must consume the
meal as well. The situation is reversed with sunflower, which
produces much less meal per unit of oil. For on-farm sunflower
processing to achieve its full potential benefits, however,
two problems must be resolved: satisfactory replacement of
soybean meal in hog rations by sunflower meal with a fairly
high oil content, and use in diesel engines of 100% sunflower
oil in a form that can be produced on-farm.

These conclusions suggest the high degree of oversimpli-
fication that has pervaded many of the statements that have
been made about the economics of vegetable oil as a diesel
fuel. It is not enough to consider only the oil and use its
current price in comparison to that of diesel fuel. One must
also take account of the other products that are produced
when the oil is extracted, and then carry the analysis back
one step to the production of the oilseed. The use of all
products must be considered, especially in comparisons of on-
farm and centralized facilities. The value and use of the
meal affects not only the effective price of the oil in both
situations, but also constrains on-farm production in a way
that does not affect centralized processing, under which the
farmer is free to purchase separately whatever amounts of oil
and meal are needed for the particular farm. Finally, any
economic analyses of this type must be regarded as closely
linked to new research results on the best form in which to
use vegetable oil, the feasibility of on-farm production of
oil in this form, the maximum proportion of vegetable oil in
the fuel, and the usefulness of farm-produced meal in livestock rations.

Acknowledgement

This work was supported by Pacific Northwest Laboratories, Battelle Memorial Institute, under the project
"Growing Oilseeds as an Alternative Fuel Source" (Contract
No. B-96221-A-Q). However, the opinions, conclusions and
recommendations are those of the author and do not necessarily reflect the views of Battelle Memorial Institute.

References

1. W. Lockeretz. "Macro-level Implications of Using Soy-
    bean Oil as a Diesel Fuel." In Beyond the Energy
    Crisis: Opportunity and Challenge, Vol. III. Pro-
    ceedings of the Third International Conference on
    Energy Use Management, Berlin (West), Oct. 26-30,
    1981. (Pergamon Press, Oxford, 1981) pp. 1667-1678.

2. R. Meekhof, M. Gill, and W. Tyner. Gasohol: Prospects
    and Implications. Agricultural Economic Report No.
    458. U.S. Dept. of Agriculture, Economics, Statistics,
    and Cooperatives Services. (Washington, D.C., June,
    1980).

3. Schnittker Associates. Ethanol: Farm and Fuel Issues.
    Prepared for the U.S. National Alcohol Fuels Commis-
    sion. (Washington, D.C., 1980).

4. Energy from Biological Processes. Vol. II--Technical
    and Environmental Analyses. Office of Technology
    Assessment, Congress of the United States, (Washington,
    D.C., September, 1980).

5. Energy and U.S. Agriculture. Statistical Bulletin No.
    632. U.S. Dept. of Agriculture, Economics, Statistics,
    and Cooperatives Service. (Washington, D.C., April,
    1980).

6. R. K. Perrin. "The Impact of Component Pricing of
    Soybeans and Milk," American Journal of Agricultural
    Economics, 62 (1980) 445-454.

7. N. J. Updaw. "Social Costs and Benefits from Component
    Pricing of Soybeans in the United States," American
    Journal of Agricultural Economics, 62 (1980) 647-655.

8. K. R. Kaufman and G. L. Pratt. "Sunflower Oil
    Production/Utilization." In Alcohol and Vegetable Oil
    as Alternative Fuels. Proceedings of regional work-

shops, sponsored by Purdue University, 1981, pp. 233-243.

9. G. Pratt, L. Backer, K. Kaufman, L. Jacobsen, C. Olson, P. Ramdeen, W. Dinusson, D. Helgeson, L. Schaffner, and H. Klosterman. "On Farm Production of Sunflower Oil for Fuel." In Beyond the Energy Crisis: Opportunity and Challenge, Vol. III. Proceedings of the Third International Conference on Energy Use Management, Belin (West), Oct. 26-30, 1981. (Pergamon Press, Oxford, 1981) pp. 1767-1774.

10. D. Helgeson and W. Schaffner. "The Economics of On-Farm Processing of Sunflower Oil," North Dakota Farm Research 39(4), (1982) 3-7, 13.

11. J. J. Bruwer, B. V. D. Boshoff, F. J. C. Hugo, J. Fuls, C. Hawkins, A. N. v.d. Walt, A. Engelbrech, and L. M. du Plessis. "The Utilization of Sunflower Seed Oil as A Renewable Fuel for Diesel Engines." In Agricultural Energy, Vol. 2, (American Society of Agricultural Engineers, St. Joseph, Michigan, 1981) pp. 385-390.

Edward S. Lipinsky, Thomas A. McClure,
Stephen Kresovich, James L. Otis,
Cynthia K. Wagner

11. Production Potential for Fuels from Oilseeds in the United States

Trends in petroleum production, refining, and utilization can be expected to exert heavy pressure on middle distillate fuel supplies in the United States during the next several decades. The major consumers of middle distillate fuels are diesel engines that operate most long-distance trucks and railroad locomotives, as well as most agricultural machinery and gas turbine power plants employed by utilities in peaking applications. The increasing number of diesel passenger automobiles will tend to increase middle distillate fuel consumption. Although synthetic fuels from coal or oil shale could close any gap between potential supply and demand for middle distillate fuels, the noticeable lack of progress in construction of large-scale synfuels facilities does not offer as much cause for optimism now as it did a few years ago. Furthermore, the fact that petroleum prices have fluctuated instead of rising steadily has increased the risk of investing in huge synfuel enterprises.

It is against this background of increasing pressure coupled with uncertainty that the Department of Energy and the Electric Power Research Institute have sponsored recent research programs that relate to middle distillate fuels (1, 2). These systems studies concern the possible use of the vegetable oils derived from vegetable biomass sources as substitutes for middle distillate fuels. The studies cited have many other ramifications that pertain to terpenoid crops and animal fats and to lignocellulose-derived fuel (3). This chapter is primarily focused on the more limited issues associated with the development of crop production systems to replace petroleum middle distillate fuels with vegetable oil-derived fuels.

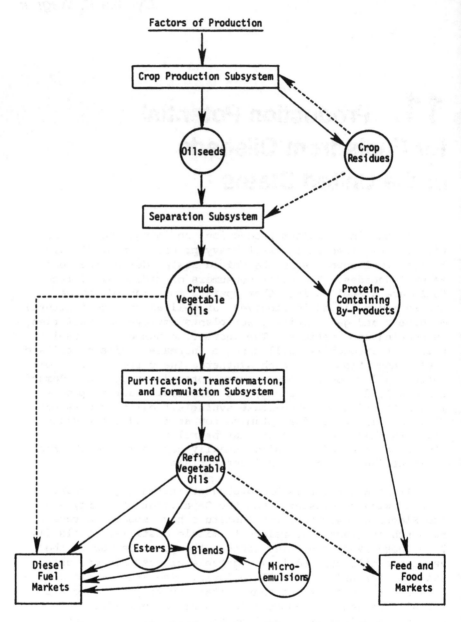

Figure 1.   Overview of vegetable oil systems.

The Vegetable Oil Fuel
Production System

As shown in Figure 1, the vegetable oil fuel system is comprised of three major subsystems. The crop production uses the factors of production (land, labor, capital, and energy) to provide an oilseed output, along with crop residues. The oilseed output in turn is the raw material for a separation subsystem that yields crude vegetable oils, plus a protein-containing by-product. The crop residues may provide either energy for the separation process or nutrients for the crop production system, depending on the degree of centralization of the separation process and soil conditions. The crude vegetable oils (triglycerides) are refined and/or transformed and formulated into middle distillate fuel products. Typical purification processes include degumming to remove phosphates, winterizing to removing to remove high melting components, and alkali refining to remove free fatty acids. The protein-containing by-products are sold to animal feed/food markets. The final vegetable oil products may be suitable for industrial or food markets, in addition to the fuel markets.

The Agricultural Fuel Market

As reported by the American Petroleum Institute (4), the overall market for middle distillate fuels exceeds 40 billion gallons per year, including 3 billion gallons of agricultural fuel. The market is divided approximately in half between transportation and nontransportation applications. The overall market is perhaps twenty times as big as the current production of vegetable oils in the United States; therefore, it is unlikely that vegetable oils will displace a large fraction of the petroleum products from this market. Nevertheless, the three billion gallon per year agricultural segment does represent an enticing target. The market also could be segmented in terms of geographical criteria to yield promising local markets. Because it is the selling price of the petroleum product delivered to the site of use that is the determining factor, vegetable oil fuels are likely to be favored in rural areas where oilseeds are grown and processed far from the main lines of petroleum distribution.

On-farm production of diesel fuel substitutes has economic implications for the individual producer that differ from those operating in the open market economy. These special considerations may render it possible for the on-farm producer to make a profit from using his own triglyceride fuel even though it is mathematically more expensive than purchased. Typical special considerations include: 1) a tax writeoff on processing equipment that would be beneficial to farmers in

relatively high tax brackets; 2) off-specification oilseeds
that could be put to good use; 3) cash outlays could be reduced because less fuel is purchased; and 4) when oilseed
prices are low, oilseeds could be converted to fuel instead
of being sold at a loss or stored at a cash cost. The avoid-
ance of cash payments for fuel and taxes appears to be the
strongest reason for not requiring equality of vegetable oil
fuel prices with petroleum middle distillate fuel prices.

Agricultural Systems Considerations

The successful development of oilseed systems to provide
triglyceride fuels depends on the choice of cropping system.
Specification of the cropping system involves the selection
of species and cultivar that largely determine the composition of the triglyceride and its by-products, the choice of
land, and the choice of the crop management system. The potential yield of an oilseed crop is the result of a number of
biological, chemical, and physical interactions. Many of
these factors can be manipulated to some extent (e.g., soil
fertility and irrigation), whereas others cannot (e.g., photosynthetic pathway). Considerations of productivity include:

- Level of crop development (amount of genetic manipu-
lation and size of the genetic base)
- Climatic, soil-related, and water-related factors
- Carbon fixation patterns (when/where the reduced
photosynthetic carbon is translocated)
- End-product composition
- Length of crop growth cycle (annual versus biennial
versus perennial)
- Level of agricultural management (extent of use of
specific fertilizers, irrigation practices and pest management practices).

Productivity is not equivalent to profit in that different
management practices and costs are incurred to maximize profit (revenues minus cost), compared with maximization of fuel
production per unit area of land or maximization of net
energy production per unit area of land. U.S. agriculture
tends to make decisions primarily on a net profit basis.

Where net profits are the primary concern, the level of
agricultural management will be adjusted to minimize the use
of hired labor and borrowed capital. This guideline gives a
great advantage to those crops that can be managed with existing agricultural equipment or with addition of minor capital
items to supplement existing equipment. Perennial plant types
may be advantageous over annuals because an effective canopy

is present over a greater period of the growing season, which
allows photosynthesis to occur for a longer time period.
However, the advantages of a perennial crop frequently are
offset by their failure to produce salable products during
the first year or years after planting.

The average and estimated potential commercial yields of
selected oilseeds are displayed in Table 1. The estimates of
potential commercial yields were based on discussions with
experts in breeding and agronomy of the selected oilseeds,
analyses of potential yields under experimental conditions,
and a knowledge of crop improvement considerations. They are
not meant to reflect absolute values but rather to present a
relative index. In addition, the confidence limits associated with potential yields of commercial oilseeds (e.g., corn
and soybeans) are much narrower than are the estimates for
the newer oilseed crops, such as the Chinese tallow tree
(Chapter 6).

Rotation Cropping Systems

Rotation cropping is defined as the repeated cultivation
of an ordered succession of crops on the same land with one
cycle requiring several years to complete. Among the promising rotation cropping systems which may evolve over the next
two decades is the combination of corn and sunflowers. This
combination could replace much of the acreage that is currently utilized for corn and soybean rotation in the Corn Belt and
the Lake States. This change, if implemented, would potentially increase triglyceride production from a given acre by
about 50 percent based on the current oil yields of soybeans
(41 gallons per acre) and sunflowers (61 gallons per acre).
However, the corn/sunflower rotation system would yield a
correspondingly smaller quantity of protein.

In the development of new rotation cropping systems, considerations must be given to economic and environmental
issues, as well as to increasing yields of vegetable oils.
The integrity of the soil and water resource base must be
maintained, especially when a leguminous crop such as soybeans
or alfalfa is replaced by a non-leguminous one such as sunflowers. Losses of prime farmland due to erosion, salinity,
and other causes must be held to a minimum.

Sequential Cropping

Sequential cropping differs from rotation cropping in
that it involves growing two or more crops in sequence on the
same field during one year. Sequential cropping patterns

Table 1. Average and Potential Commercial Yield of Selected Oilseeds in the United States

Oilseed Species	Seed yield (lb/acre)		Oil yield (lb per acre/gal per acre)				Growth Conditions
	U.S. Average	Potential	U.S. Average Lb.	Gal.	Potential Lb.	Gal.	
Castorbean (Ricinus communis)	848	3,400	382	48	1,342	170	Irrigated
Chinese Tallow Tree (Sapium sebiferum)	---	----*	---	--	---	--*	Dryland
Cotton (Gossypium hirsutum)	791	1,700	127	16	306	40	Irrigated
Crambe (Crambe abyssinica)	1,000	2,100	350	45	735	100	Irrigated
Corn (high oil) (Zea mays)	---	5,300	---	--	532	70	Dryland
Flax (Linum usitatissimum)	709	1,600	253	33	676	90	Dryland
Peanut (Arachis hypogaea)	2,122	4,600	673	87	1,458	190	Dryland
Safflower (Carthamus tinctorius)	1,495	2,200	493	64	792	100	Dryland
Soybean (Glycine max)	1,767	3,000	316	41	527	70	Dryland
Sunflower (Helianthus annuus)	1,182	2,200	473	61	880	110	Dryland
Winter Rape (Brassica napus)	---	2,400	---	--	958	130	Dryland

Source: Lipinsky, et al, 1981 (Reference 1).
*A seed yield of 11,200 pounds per acre has been reported by W.M. Potts in the Chemurgic Digest 5, 373, 375 (1946), and an oil plus tallow content of 44.2% has been given by W.M. Potts and D.S. Bolley in Oil and Soap 23, 316-318 (1946). This is equivalent to approximately 670 gallon/acre. However, in a managed but dense stand, the Chinese tallow tree seed yield probably would be substantially lower than 11,200 pounds per acre, but still higher than from conventional oilseed crops in the United States.

should provide additional means for increasing oilseed pro-
duction from current cropland. On a large scale, the most
promising cropping pattern is the growing of oilseeds following the production of winter small grains. Examples include:
- Sunflower following wheat in the northern plains
- Soybeans following wheat/barley in the Southeast,
and southern corn belt
- Winter rape following corn/cotton/sorghum in the
Southeast and southern plains
- Peanuts following wheat/rye in the Southeast.
In addition, the sequential cropping safflower following
winter vegetables in the mountain and Pacific states might
increase production.

The economics of double cropping wheat and soybeans in
Ohio has been studied by researchers at The Ohio State
University. Although soybean yields are lower than the
national average due to a short growing season, a soybean
yield of twenty bushels per acre should be possible in an
"average" year, at least in the southern half of Ohio.
According to Triplett and coworkers (5), double cropping
wheat and soybeans would allow the grower an estimated return
over variable costs of $158 per acre, compared with $95 per
acre for growing only wheat, assuming a wheat yield of 45
bushels per acre. When land and other fixed costs are included, return for double cropping wheat and soybeans is $9
per acre, compared with an estimated loss of $28 per acre
for wheat alone. It is quite possible that double cropping
of wheat with sunflowers would provide a higher return. The
key assumptions underlying these estimates are: wheat yield
of 45 bushels per acre, soybean yield of 20 bushels per acre.
The fixed costs for production of wheat/soybeans is estimated
at $149 per acre vs. $123/acre for wheat alone.

The required crop management skills are much greater
under sequential cropping systems than under rotation cropping systems. Water availability is critical, for example.
Also, the sequential cropping systems demand timely and careful management, including early harvest, use of short-season
cultivars, narrow rows for double cropped oilseeds, selection
of herbicides based on planting methods and crop sequence,
and consideration of no-tillage planting of the summer oilseed crops.

Even with proper management, sequential cropping entails
greater risks which often are beyond the control of the grower. For example, soybeans planted in Ohio in early-to-mid-
July following wheat harvest are subject to potential yield
and quality reduction occurring from the early frost which

could occur in mid-September, even though the normal date of
first killing frost is in October.

Intercropping

The growing of two or more crops simultaneously in the
same field is designated "intercropping". This cropping
strategy may prove useful in two quite different strategic
situations. In the first, a second crop is planted and
allowed to germinate during the time that the first crop is
reaching maturity. The disadvantages of the mature crop
shading out the second crop are avoided because the initial
crop is harvested before the second crop needs much sunlight.

Agroforestry is an example of a second type of intercropping system (6). A perennial plant that may require years
to attain harvestable status is intercropped with a short
season crop. This cropping system permits some cash revenue
from the land during the early stages of growth of the perennial crop. Intercropping of the Chinese tallow tree,
currently uncultivated in the U.S., with peanuts would be an
example of such a system.

Upgraded Cropland Systems

Rotation cropping, sequential cropping, and intercropping generally make use of the best quality land because they
emphasize maximization of output and profits. An entirely
different strategy for obtaining triglyceride fuels is to
leave the present food and fiber system intact while upgrading pasture, range, and forest lands of sufficient quality
into the production of additional oilseeds. This type of
cropping system could add to both the fuel and food supply of
the nation. As the discussion presented below indicates, the
aggregate acreage that is potentially available might suffice
to provide enough oilseed raw material to permit the production of one billion gallons per year of triglyceride fuel from
these upgraded croplands.

A National Agricultural Lands Study, published in 1980,
in which several government agencies participated, provided
information on land available for agriculture in the nation's
ten farm production regions (7). About 127 million acres of
land currently not cropped were indicated as having a "high
or medium potential" for use as croplands. Most of this
acreage now is in forest, pasture, or range. According to
the 1977 survey, 36 million acres of the 127 million acres
have a high potential for conversion to cropland, while 91
million acres have medium potential. A "high potential"

rating indicated the presence of favorable cropland characteristics such as adequate moisture supply, soil temperature,
water permeability and slope. A "medium potential" rating
also required somewhat favorable characteristics, but the
cost of conversion of medium quality land was expected to be
higher than for land given a higher potential rating.

Figure 2 indicates existing cropland and land area designated as having high or medium potential for conversion to
croplands, by state, for the year 1977. The relationship of
the existing cropland to the total cropland base is also indicated in Figure 2. Thus, Iowa is making use of about 90
percent of its potential cropland but North Carolina and
Georgia are using only about half of their potential cropland. Unfortunately, the quantity of land potentially available for conversion of cropland changes constantly as underutilized forest land is planted in high quality tree crops,
prime farmland is converted to subdivisions of cities, etc.
Therefore, this 127 million acre estimate can only be considered a rough estimate.

When potential cropland presently in forests, pasture,
or range is converted into the land resource base for an oilseed cropping system, yields lower than for existing cropland
can be expected. Furthermore, a considerable fraction of the
potential cropland will never be converted into actual cropland due to adverse ownership patterns. For the purpose of
estimating potential vegetable oil production from such land,
the team conducting a recent DOE study assumed 20 percent of
the cropland could be employed for oilseed production and
that reductions in yield would average about 25 percent (1).

With these assumptions, the estimates in Table 2 result.
All of these yields are substantially less than the potential
yields estimated in Table 1 because of lower quality land and
less intensive crop management practices. The Pacific,
Mountain, and Northeast farm production regions are not included at all in this estimate because they either 1) do not
have large quantities of land potentially convertible to
cropland, or 2) do not produce significant quantities of
oilseed. However, these lands could be available to produce
these crops, if necessary, in addition to other oilseeds
that might be especially suitable to these regions (e.g.,
winter rape or safflower).

Using the assumptions noted above, the additional potential vegetable oil production in the seven production regions
might be as high as one billion gallons. The largest expansion of oilseed production would be in sunflowers and peanuts.

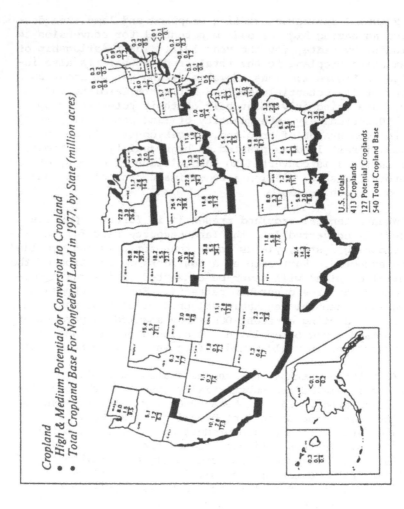

Figure 2. Cropland and potential acreage for conversion to cropland, 1977, by State (7).

Table 2. Possible Reallocation of Land Use to Achieve Vegetable Oil
Production of One Billion Gallons

Regions	Crops	Land (Million Acres)	Vegetable Oil Yield (Gal/Acre)	Production (Million Gal/Year)
Northern Plains	Sunflowers	4.2	45	200
Southern Plains	Sunflowers	4.2	45	200
Lake States	Soybeans	3.3	30	100
Corn Belt	Soybeans	3.3	30	100
Delta States	Soybeans	2.3	30	70
Southeast	Peanuts	4.3	70	300
Appalachian	Soybeans	3.3	30	100
Total		24.9	--	1,070

Source: Lipinsky, et al, 1981 (Reference 1).

Table 3. Long-Term Potential for Production of Vegetable Oil Fuels from Oilseed Cropping Systems in the United States

Cropping	Basis for Estimate		Estimate(a) Billion Gal/Yr	Typical Number of Processing Plants Required			
	Million Acres	Yield Gal/Acre		On-Farm(b)	Cooperative	Small Commercial(d)	Large(e) Commercial
Rotation	20	60	1.2	170,000	7,500	200	20
Sequential	25	40	1.0	143,000	6,250	167	17
Upgraded Cropland	20	40	0.8	114,000	5,000	133	13
Chinese Tallow Tree	6	---(f)	1.8	243,000	10,625	283	28
Total	71	---	4.8	670,000	29,375	783	78

(a) Estimate is for a year that is at least 15 years after onset of a chronic emergency serious enough for a concerted effort to be launched to commercialize these crops.
(b) A small facility to produce about 7,000 gallons per year.
(c) A local cooperative unit to produce about 160,000 gallons per year.
(d) A small commercial facility to produce about 6 million gallons per year.
(e) A commercial facility to produce about 60 million gallons per year.
(f) See footnote to Table 1. Although approximately 670 gallon/acre can be calculated from published literature, it seems likely that actual yields from managed but dense stands will be significantly lower, but still higher than from conventional oilseed crops in the United States.

Source: Adapted from Lipinsky, et al., 1981 (Reference 1).

Sunflower acreage would be 2 to 3 times as large as current production while peanut production would more than double.

The estimates shown in Table 2 are not projections of what should occur or what is likely to occur. Rather, the estimates are intended to provide some perspective on the potential contribution that oilseed crops that have already been commercialized could yield when matched with potential land availability. Important agricultural considerations that would need to be addressed before undertaking such a program of oilseed crop production would be the availability of sufficiently large contiguous land areas, marketability of by-products, and availability of economical oilseed processing equipment. In addition, the use of higher value land in sequential or rotation crop systems might prove to be superior to conversion of marginal land to production of oilseeds. Alternatively, entirely new oilseed crops, such as the Chinese tallow tree, might prove to be more profitable than a marginal land system for conventional oilseeds (See Chapter 6).

Long Term Potential for Vegetable Oil Fuel Output

The cropping systems described above could be employed to produce significant quantities of vegetable oils and chemical feedstocks if a chronic emergency ever arose that was considered serious enough for a concerted effort to be launched to commercialize these crops as petroleum replacements. As shown in Table 3, each of the cropping systems discussed in this chapter has the potential to make a contribution in the billion gallon per year range. The Chinese tallow tree is illustrative of new oilseed crops that could be introduced onto land that is not crucial for food or high quality forest products. Other oilseeds could be substituted, with corresponding changes in the acreage required to obtain one billion gallons.

Also shown in Table 3 are estimates of the numbers of facilities required to process this quantity of additional oilseed production. If only small on-farm units were employed, more than 0.5 million facilities would be required. This large number of facilities arouses skepticism because an unusually large number of farmers would have to be interested in processing these crops and have the credit required to construct the facilities. Offsetting this disadvantage would be the economies of scale for the mass producers of the processing units. Local cooperatives and small commercial units would be required in correspondingly smaller numbers. Fewer

than 100 world scale oilseed processing facilities could provide the estimated 4.7 billion gallons of triglyceride fuel
per year. If the chronic emergency situation did arise, one
would expect a mixture of size ranges to be successful in
providing the productive capacity for the triglyceride fuels.
The size of facility would depend on local demand, the properties of the oilseed to be processed, and the credit situation.

Economics of Vegetable Oil Recovery

An economic analysis has been carried out of peanut and
sunflower oil production using on-farm, cooperative and small
commercial units. Under normal (1981) economic conditions,
the cost per gallon of diesel oil from commercial units is
estimated at $1.76 - 1.77 for both crops. Under emergency
conditions of zero profit, costs per gallon of $1.63, $2.20
and $1.85 respectively, are obtained for on-farm, coopera-
tive and small commercial unit production of sunflower oil;
corresponding costs for peanut oil of $1.81, $2.08 and $1.68
were obtained (1). Conventional commercial oilseed processing facilities operate at ten to fifteen times the scale of
the small, local commercial facilities considered in the DOE
emergency fuel program (1). The economies of scale of large
plants are so great that they could make vegetable oil available at about $1.50 per gallon (1981 prices) which is much
less than the small plants can accomplish. Smaller scale
facilities made in factories using modular design might be
more competitive by pitting production line economies against
scale economies. However, this concept has not been explored
enough to reach a conclusion. Until new technology can be
developed that favors small-scale processing, the balance of
power appears to be with the 60 million gallon per year
scale. The economies of scale of conventional, commercially
sized operation are clearly evident. The low cost of on-farm
operation compared to the cooperative unit can be attributed
to use of some existing or used equipment, low cost of labor
for operation and plant construction, and low overhead.

The Food Versus Fuel Question

Since the early 1970's, when the Unites States' energy
problems were brought to the forefront, the issue of utilizing agricultural resources for food or fuels has been debated.
Strictly speaking, it is not so much a matter of utilizing the
agricultural resources per se for fuel, as it is a question
of using finite supplies of land and water to produce fuel.
When petroleum prices were relatively low, i.e., prior to the
1973 OPEC oil embargo, agricultural products were far too high

priced to be considered as energy feedstocks. However, as
energy and petrochemical prices have risen, agricultural commodities have become increasingly attractive as alternative
energy and chemical feedstocks, adding pressure on food
supplied and prices. The competition for land and water resources presents critical choices to policy makers to inter-
vene in free markets, either to hasten or delay the point at
which the energy-food interface is joined. The choices for
each country are weighted by the relative scarcity of fuel
sources, especially petroleum, and domestic production (8).

Up through 1980, it appeared that domestic energy supply
problems would be more severe than food shortages during the
1980's. Over the long run, energy markets may experience
greater increased demand and an earlier supply plateau than
is anticipated for food markets. If this result occurs, crops
that are usually grown for food applications will be purchased
in significant quantities for energy applications. Energy
crops and food crops also will compete directly for land,
water, and other factors of production. This diversion of
food crop production to satisfy energy demand will lead to
higher prices in food markets due to the price elasticity
effect. The magnitude and social impact of these food price
increases will be a major policy issue.

Rask (8) indicates some positive and negative effects
for the United States for enacting policies that favor energy
as opposed to food production. The positive effects include:
1) in the long run, more stable supply of liquid fuels, particularly important from a national security perspective; 2)
an improved balance of payments situation; 3) use of energy
crops as a short-run bridge to the use of shale oil and coal
liquefaction; 4) increased income and employment in rural
areas; and 5) substitutions for present "surplus food production" programs (i.e., transfer of present subsidies for income maintenance, set-aside acres, support prices, etc. to
provide incentives for crop production).

Rask's negative effects for intervening to promote energy
production are: 1) higher energy and food prices, at least in
the short run; 2) a smaller "food reserve" for the world in
times of food production shortfalls; and 3) a reduction in
food exports that might reduce foreign exchange earnings.

The effect of decontrolled oil prices presumably will be
to stimulate energy conservation and the use of renewable
resources. Following this reasoning, production of oilseed
crops as diesel fuel extenders will be stimulated by the rising oil prices.

The National Agricultural Lands study (7) emphasized
that there is no free land from which to withdraw croplands
to replace those being lost to other uses. New croplands
are available only at various costs to other segments of the
national economy. These costs may include red meat production, dairy products, wood products, the expense of converting
from current use of cropland, and significant changes in
environmental and wildlife values. Land conversion from a
dwindling land base should be made in full recognition of
these tradeoffs and in concert with them (7).

Crosson (9) discusses the effect of technological change
upon the demand and supply of cropland. Improvements in technology (e.g., improved plant cultivars, control of diseases
and pests, etc.) lower production costs per unit of output,
and consequently prices, stimulating demand for crops and the
land to produce them. Technological improvements may be land-
saving, land-using, or neutral. Land-saving technologies,
that reduce the amount of land required relative to other inputs, result in a reduced rate of increased demand for land
than if the improvements are land-using or neutral. Significant land-saving improvements may offset the effect of
higher crop demand, resulting in a net decline in the amount
of land and crops even though crop production is higher. For
example, the use of minimum tillage systems reduces erosion
and permits the cropping of land unsuitable for conventional
row crop systems. Another more recent technology, not yet
fully commercialized, involves breaking the lignocellulosic
complex in wood and cereal straws so that lignin, cellulose,
and hemicellulose can be obtained separately. This technology reportedly results in a roughage substitute with improved
digestibility and feed efficiency for cattle (10). Such a
technology has the potential to reduce the land required for
conventional forages (i.e., hay) now fed to cattle. Thus,
land formerly devoted to hay production might be transferred
into production of oilseed crops to produce vegetable oils.

In summary, even though the United States has a fixed
quantity of land for food and fuel production, technological
advances, economics, and environmental considerations will
dictate whether or not new crop lands will be developed to
produce vegetable oils as potential replacements for diesel
fuels.

Conclusions

If or when an emergency of sufficient magnitude and duration arises for the United States to undertake a massive
energy crop production program, there appears to be sufficient
land of the qualities needed for the various oilseed crops to

permit the production of one billion to five billion gallons
per year of vegetable oil fuels.

The effect of such an energy program on food availability
is not expected to be adverse because the oilseeds contain
both vegetable oil and protein. However, protein prices might
become greatly depressed, due to the huge quantities of additional protein that would be produced.

The production of one billion to five billion gallons of
vegetable oil fuels can be considered as a significant contribution to the agricultural use of diesel fuel, even at the one
billion gallon scale. At the five billion scale, the vegetable oil fuels would be making a significant contribution to
the overall consumption of middle distillate fuels in the
United States. Thus, vegetable oils might account for over
30 percent of the agricultural middle distillate market and
approximately 10 to 15 percent of the total U.S. diesel fuel
market.

Most of the candidate oilseeds are ones that some farmers in the United States know how to grow. The major problem
in mounting a vegetable oil fuel production program would be
to spread this knowledge to more farmers and to modify the
oilseed crops to meet fuel rather than food requirements.
Both plant breeding and agronomic practices development need
to be applied to sequential cropping systems to make possible
the one billion (or more) gallons which appear possible to
produce.

The Chinese tallow tree is the major exception to the
statement that most candidate oilseeds are well known. This
crop needs development in both breeding and agronomic
(silvicultural) practices, a time consuming and expensive
endeavor. The initial results indicate that such a development could be highly rewarding.

The agricultural subsystems are not likely to be the major barriers to commercialization of vegetable oil fuels.
Satisfactory processing of the fuels in facilities that are
economical to purchase and operate is a major barrier. The
other major barrier is the development of vegetable oil fuels
(neat, blends, or microemulsions) that can be used in existing
diesel engines. The manufacturers of diesel engines are also
challenged by vegetable oil fuel properties to develop engines
that are more readily compatible with middle distillate fuels
containing triglycerides.

The food-vs.-fuel controversy is not a single issue that
can be settled, once and for all time. There may be countries

in which biomass fuel programs could cause soaring food prices and inflation. The United States is at the opposite extreme with respect to oilseeds; depression and low prices are experienced by oilseed growers because the conventional food and industrial markets for oilseeds are insufficient in size. A large fuel market for oilseeds could increase the attractiveness of oilseed production and processing, making both fuel (vegetable oil) and food (protein-rich oilseed meal) more available.

Acknowledgment

The authors gratefully acknowledge support from the Department of Energy's Biomass Energy Technology Division and the Office of Vehicle and Engine Research and Development for the research on which this chapter is based. The Electric Power Research Institute also provided project support in this general area. Helpful discussions with Dr. Marilyn Ripin and Dr. Eugene Eckland of DOE and with Dr. Stephen Kohan of the Electric Power Research Institute provided a supportive basis for the investigation on which this chapter is based. The editorial and secretarial support of I. R. Morgan and J. P. Carlyle are greatly appreciated by the authors.

References and Notes

1. E. S. Lipinsky, T. A. McClure, S. Kresovich, J. L. Otis, C. K. Wagner, D. A. Trayser, and H. R. Appelbaum, "Systems Study of Animal Fats and Vegetable Oils for Use as Substitude and Emergency Diesel Fuels", Phase II Final Report from Battelle Memorial Institute to U.S. Department of Energy, October 30, 1981.

2. E. S. Lipinsky, D. A. Ball, and D. Anson, "Evaluation of Biomass Systems for Electricity Generation", EPRI AP-2265, Electric Power Research Institute, Palo Alto, February, 1982.

3. For more detail, the interested reader is referred to (1) and (2), to the Proceedings of the International Conference on Plant and Vegetable Oils as Fuels, 1982, sponsored by the American Society of Agricultural Engineers, and to the forthcoming Proceedings of the 1983 International Solar Energy Society Conference entitled "The Renewable Challenge".

4. Anon., "Basic Petroleum Data Book", American Petroleum Institute, Washington, D.C., Volume II, Number 2, Section VII, Table 11a, May, 1982.

5. G. Triplett, "Weed Control for Double Cropped Soybeans Planted with the No-Tillage Method Following Small Grain Harvest", Agronomy Journal 70, 1978, p. 577, and subsequent 1980 Ohio Farm Enterprise Crop Budgets.

6. W. E. Raitanen, "Energy Fibre and Food: Agriforestry in Eastern Ontario", 8th World Forestry Congress, Jakarta, Special Paper FFF/7-16, October 16-28, 1978, 13 pp. (mimeo).

7. Robert Gray, National Agricultural Lands Study, Interim Report #5, American's Land Base in 1977, USDA, Washington, D.C., December 1980.

8. N. Rask, "Agricultural Resources for Food or Fuel: Policy Intervention or Market Choice", Mershon Quarterly Report, The Ohio State University, Volume 5, Number 2, Winter, 1980.

9. P. Crosson, "Agricultural Land Use: A Technological and Energy Perspective", in Farmland, Food, and Future; Soil Conservation Society of America, 1979.

10. Pro-Cell Nutritional Report, A Stake Technology Processed Product; W. J. Esdale, Stake Technology, Ltd., Nepean, Ottawa, Canada, August, 1979.

Sandra Lee Mathieu, Eugene B. Shultz, Jr.

12. Oilseeds as Lighting and Cooking Fuel for Developing Nations

Introduction

The heating values of oilseeds are similar to those of charcoals, and heating values of most seed oils are nearly that of kerosene (1). Therefore, if oilseeds could be processed at village level in developing countries, inexpensive renewable energy under local control might be provided for rural areas to replace scarce and expensive charcoal and kerosene.

Unutilized oilseeds exist in many developing countries. For example, it has been estimated that five known and unutilized oilseed plant and tree species in India produce over 6 million tons of seeds per year (2). A similar study conducted in Pakistan has identified 25 plant and tree species that produce large amounts of unutilized seed fat annually (3). Further, the possibility exists that inedible oilseeds might be grown for fuel on marginal lands with minimal competition for food production (see Chapter 2).

The successful utilization of village-processed oilseeds to provide lighting, cooking and heating fuel in rural areas might help in dealing with at least three major problems facing most developing countries: (a) high petroleum import bills, by reducing the need for petroleum-based lighting and cooking fuels such as kerosene and butane, (b) the fuelwood crisis, by partially replacing wood and charcoal with oilseed fuelcakes, and (c) a high rate of migration to overcrowded urban areas, by creating new oilseed processing industry and employment opportunities in rural sectors.

Oils extracted from oilseeds have probably been used as illuminants for thousands of years. Further, it is

likely that the remaining press-cake has been occasionally
utilized as a firewood substitute. However, use of seed
oils for lighting appears to have fallen out of favor with
the coming of the petroleum era in the 19th century, and
the use of press-cakes for cooking has never been very
extensive. Consequently, the present contribution of
oilseeds and seed oils to household energy in developing
countries is probably insignificant (4).

In this chapter we will explore the possibility that
seed oils might be revitalized as lighting fuels in simple
lamps in rural areas to replace expensive imported kerosene. Kerosene has technical characteristics that make it
an easier fuel to use for lighting than seed oils, but
because the price of kerosene has increased greatly during
the last decade it may be timely to re-evaluate the cost
of its advantages.

Further, we will examine the potential for oilseeds
as fuel for cooking, by way of combustion of cakes or
pellets of ground or chopped whole seeds ("fuelcakes") as
wood or charcoal substitutes. To the best of our knowledge, the fuelcake concept has not previously been prac-
ticed or studied. Charcoal is a popular fuel in many
developing countries, but its production from wood typically results in the loss of about half the original
heating value of the wood. About four pounds of wood are
required to produce one pound of charcoal (5). Therefore,
because oilseeds and charcoals have similar energy densities, one pound of oilseeds with about 50 percent oil is
the approximate energy equivalent of four pounds of wood
when the wood is used for charcoal manufacture. The
implication is that if satisfactory technology can be
developed for using oilseeds as cooking and heating fuels,
oilseeds might compete in the market system with charcoal
and help alleviate the loss of wood to charcoal production.

Seed Oils for Lighting

Background

During the early to mid-19th century when fatty oils
were in common use as illuminants, the best lighting oil,
sperm whale oil, sold for as much as $2.50 per gallon.
Equally high were the prices of vegetable oils also considered to be of top quality for lighting, such as mustard
seed, rape or colza, and various nut oils (6). When
kerosene was introduced, it offered a low-viscosity fuel
with good combustion and storage properties, all at a

fraction of what the quality fatty oils had cost. Kerosene's market position was further strengthened by the advent of more modern lamp designs, compared to seed oil lamps. Although round wicks and glass chimneys to improve the lighting ability of a flame had been known and proved for some 60 years before the introduction of kerosene, it was not until the widespread adoption of kerosene that these and other design improvements became common (7).

In many developing nations, another factor helped bring about the widespread utilization of kerosene as an illuminant in recent years. Faced with devastating deforestation problems, many countries perceived the use of kerosene as a solution. Programs were established to make subsidized kerosene widely available (8). Many such programs that began prior to the 1973 oil embargo and the soaring price of petroleum have since been eliminated or severely reduced. However, the encouragement of use of kerosene may still be felt because most of the lamps in use are suitable only for kerosene. The physical and chemical characteristics of seed oils and kerosene differ markedly. Therefore, seed oils cannot be used in kerosene lamps. The few lamps now available that were designed for vegetable oils are copies of antiquated oil lamps which do not incorporate many of the design improvements and new knowledge of illumination usually found in simple kerosene lamps. Consequently, even where seed oils may have been cheaper, there have been no modern lamp designs to utilize these alternate fuels effectively.

Seed Oil Properties

Seed oils have higher values of viscosity, surface tension, and smoke, flash and fire points than kerosene. The result is that seed oils do not travel through a wick with facility, and are much less volatile than kerosene. Viscosity and surface tension are increased by increases in molecular chain length. Unsaturated oils tend to be less viscous with higher surface tension in comparison with oils that are more saturated. The presence of free fatty acids (FFA), because they are more volatile than triglycerides, tends to lead to a lowering of smoke, flash and fire points (4).

Other factors that influence oil characteristics include species, climate, and seed ripeness (9). For example, oilseeds that produce relatively unsaturated oils are more typical of temperate climates, while relatively saturated oils tend to predominate in tropical regions.

Seed oils of the same species have been found to be more unsaturated the farther north the seeds are grown. Further, because fatty acid composition will usually change during seed development, timing of the harvest may affect seed oil composition. Saturated acids tend to dominate early in seed development, and unsaturated acids, such as oleic and linoleic, usually are formed later. Early harvesting may also affect the percentage of oil in the seed; in many oilseeds the bulk of the oil is deposited late in the ripening process.

Lamp Design

Since the viscosity of seed oils is much higher than that of kerosene, capillary action through the wicks is limited in practice to only a few inches. If oil were to be conducted to the flame from a reservoir, the oil would have to be within a few inches of the flame even as the reservoir was being depleted. There have been four approaches to solution of this problem: (a) modification of the shape of the reservoir, (b) use of tilting (canting) lamps, (c) use of gravity to feed oil to the flame from a higher reservoir, and (d) use of a small pump to deliver oil from a lower reservoir. The most popular modifications of reservoir shape included a conical form in which the bulk of the oil was stored close to the flame, a simple approach. Also simple were some tilting lamp designs in which the oil reservoir pivoted as oil was consumed, maintaining the oil level at a constant distance from the flame. The other two approaches, gravity feed and pumps, appear to be complicated and unlikely to prove cost-effective in village service (4).

Optimum light is produced if ample oxygen is available and if the flame temperature is high. Heat losses from radiation and convection reduce the amount of illumination that can be obtained from a flame. Globes and chimneys help to reduce these losses, and can be utilized to preheat the air coming into the flame as well as to raise the air temperature around the flame. Oxygen supply and, therefore, candlepower can be dramatically improved by two simple design improvements: use of a chimney and a hollow or thin wick. The first device to incorporate this combination of features was a gravity-fed vegetable oil lamp known as the Argand burner (6, 10), introduced in 1783 and rated at 10 candlepower. The chimney created a draft to increase the rate of oxygen supply, and also served to isolate the flame from cross drafts. The hollow wick served to promote diffusion of oxygen to the oil from the inside as well as the outside.

In experiments conducted on seed oils by one of us (SLM), there were considerable differences exhibited by different types of wicks, in the same lamp, using the same oil. Flame sizes and characteristics were affected by choice of wick. Flame heights between 3/4 and 4 inches were recorded. Some flames were relatively smokeless and others were smokey with production of particles of free carbon. The wick materials tested included jute, sisal, and cotton (4). How the wicks are constructed, as well as their shape, may also influence flame characteristics. In the case of kerosene, Elliott has reported that braided cotton wicks were more effective than twisted cotton fibers, and that round wicks produced up to one-third more light than flat wicks (11).

Another way to design for high candlepower is to use several flames in close proximity to each other. According to Thwing (6), two flames give 2.25 times the amount of light that a single flame provides, and three flames give 3.86 times the amount of light.

The possibility of incorporating incandescence into seed oil lamp design should also be considered. A standard kerosene lamp may only be able to produce from 10 to 20 candlepower, but it is reported that a kerosene lamp vaporizing its fuel to heat a mantle containing rare earth oxides can achieve 75 candlepower (12). The use of incandescent mantles is credited with increasing the luminosity of a flame by a factor of six over previous lighting technology (13). Unfortunately, the principles of this more sophisticated form of lighting were not developed until long after seed oils had faded from popular use. As a result, there evidently has never been any attempt to devise such oil lamps. Lamps using lime, salts, or other locally-derivable materials as the radiators might be feasible for village-level production and utilization. It might also be possible to devise an oil lamp to utilize commercial gas mantles.

Choice of Oil

Experimentation is required to determine what types of seed oils work best with what types of lamps. It is evident that there is much diversity in the burning characteristics of seed oils. It seems likely that low viscosity and high surface tension will be desirable to facilitate movement of oil in the wick. However, if this is accompanied by too high a degree of unsaturation, it may lead to problems. Spontaneous auto-oxidation and polymerization of highly unsaturated oil stored in the

reservoir would make cleaning difficult. Polymerization
of oil in the wick might lead to a charred wick and a
smokey flame.

Calculations performed according to the rules of
Steinmetz for estimating the luminosity index show that
for seed oils, this index will probably range between 1.7
and 2.2 (4). In comparison, kerosene has a luminosity
index of 2.2 (14). These rules also state that smoke
cannot be eliminated from illuminating fuels with indices
of less than 2.0, without the direct addition of oxygen.

Village Practice

For seed oils to be widely accepted as lighting
fuels, simple and efficient village oil lamps must be
developed to accomodate the conditions found in rural
areas of developing nations. The design of these lamps
for village use imposes certain restrictions. A lamp must
be simple to build and repair, and the cost should not
exceed a few days income for an average villager. The
lamp should also be designed to sustain abusive treatment
so that repairs can be minimized. Inexpensive local
materials of construction are preferable, and the choice
of materials should be consistent with the environment in
which a lamp is to be located. For example, if a lamp is
to be used in a coastal village, materials to resist
corrosion by salt would be indicated. The lamp should be
almost fool-proof in operation, perhaps even requiring no
written instructions. The oil reservoir should be closed
to prevent accidental spilling, but the reservoir should
be easy to open and clean. The lamp should be capable of
burning for several hours unattended. These features have
not been incorporated in previous seed oil lamps; however,
there seems to be no evidence to suggest that it is impossible or impractical to do so.

Oilseed Fuelcakes for Cooking
or Heating

Background

Although whole oilseeds have occasionally been strung
together on sticks and used for lighting (15), we are not
aware that they have been used for cooking or heating
fuels, either whole, or through the preparation of fuelcakes such as can be made by crushing the whole seed and
tightly packing the crushed material without removal of
any part (4). We have found that a common hand-cranked
food grinder is adequate for this purpose; other methods

can be used. We have found that crushing and packing to
make a cake or pellet is necessary to avoid the rapid,
smokey, and sometimes explosive combustion that is typical
of many whole seeds.

Laboratory Studies

Fuelcakes of several sizes and shapes have been made
from Chinese tallow tree seeds by Mathieu (4) and burned
successfully in the laboratory, including round, long and
narrow, and tall, candle-like shapes. In addition, pellets have been molded and tested successfully. All could
be lit with a single match, and all burned steadily and
smokelessly in the open, without the presence of stove or
cooking utensil. Only the candle-shaped fuelcake failed
to burn completely.

Another series of tests was conducted on a variety of
ground and pelletized oilseeds, and hulls, to measure (a)
the time required to ignite, (b) the burning time (0.5
gram samples), and (c) the average flame height. The
results, given in Table 1, indicated that all samples were
ignitable and that combustion could be sustained (4).

A third series of tests was carried out with 5 gram
samples of "simulated" oilseed pellets as fuels for the
boiling of water. Flame height and water temperature were
measured. Materials tested included: (a) pellets of
sawdust and sunflower oil with varying percentages of oil,
and (b) similar pellets that included ground hulls. In the
former tests, the effects of oil content were studied from
25 to 60 percent oil, in 5 percent increments. With increases in oil content, the pellets burned with greater
flame heights. Those flame heights above 10 to 12 centi-
meters were smokey; at lesser flame heights the flames
were smokeless. Sustainable combustion was difficult
below about 35 percent oil. Those pellets with greater
than about 50 percent oil burned rapidly with a higher,
smokier flame. Regression analysis indicated that a
transition to a smokey flame occurred at 45 to 55 percent
oil content (4).

The pellets with small additions of finely-ground
hull were made from 50 percent oil mixtures of sawdust and
oil. Distinct differences were noted, in comparison with
tests on pellets of the same composition but with no hull
content. The presence of even small amounts of hull was
effective in slowing the combustion process. Finally,
fuelcakes were made from a number of oilseeds and subjected to the same water-boiling tests given to the

Table 1. Results of ignition tests on ground and pelletized oilseeds

Seed*	Ignition time, seconds	Burning time, seconds	Average flame height, inches
Coconut (k)	3	120	3
Chinese Tallow (w)	7	121	1.5
Chinese Tallow (h)	16	30	1
Chinese Tallow (k)	9	140	3
Peanut (k)	37	83	1
Hazel Nut (k)	14	202	2
Hazel Nut (h)	20	60	0.5
Almond (k)	15	215	1.5
Almond (h)	23	23	1.5
Orange (w)	50	85	2.5
Pecan (k)	17	104	5
Pecan (h)	29	40	0.5
Brazil Nut (k)	26	80	–
Brazil Nut (h)	39	45	0.5
Pumpkin (k)	22	99	0.25
Walnut (k)	24	99	3
Walnut (h)	21	44	2
Sunflower	35	100	3
Tung (k)	22	107	3
Tung (kh)	21	74	1
Tung (h)	35	20	–

k = kernel   h = hull   w = whole seed

* Seeds are ordered according to their estimated seed oil Iodine value, beginning with the lowest value. Source: Reference 4.

simulated oilseeds made of sawdust and oil. Results were remarkably similar for real and simulated oilseeds of the same or similar oil content (4).

Although these are preliminary laboratory tests that need to be confirmed and extended, it seems clear that there is a practical upper limit to oil content of an oilseed fuelcake, above which a tall, smokey and inefficient combustion process will ensue. This limit may be in the range of 45 to 55 percent oil. Further, a practical lower limit of about 35 percent oil may be likely, below which it may be difficult to support continuous combustion. The presence of ground hull, normally included if the whole seed is crushed and used, is apparently advantageous in moderating the rate of combustion. In short, the results of the laboratory studies suggest that oilseeds might become practical household fuels for cooking and heating if they are ground or crushed, mixed if necessary with additional ground hull or similar woody material to bring the oil content down to the 35 to 45 percent range (if oil content is higher), then fashioned into cakes or pellets and burned in place of charcoal or wood.

Cookstove Tests

Our associate, Edgar W. Schmidt, has carried out tests of oilseed fuelcakes in a portable low-mass stove constructed from a five-gallon bucket, similar to an East African jiko (16). Comparisons were made with wood and charcoal as fuels in the boiling of one liter of water. Wood and charcoal were smokeless fuels; some smoke was observed in the burning of oilseeds, but measured thermal efficiencies were apparently high and the general usefulness of oilseeds as a fuel for low-mass cookstoves seemed to be confirmed. Schmidt concluded that at this stage of oilseed fuelcake development the amount of smoke formation probably would prevent acceptance of oilseeds as an indoor fuel without a chimney, in some cultures. Several new lines of research were proposed, including the use of additives to moderate the rate of combustion.

Estimation of Fuelcake Demand

Results of preliminary tests we have performed on the cooking of the traditional boiled meal, irio, of the Kikuyu tribe of Kenya, East Africa, suggest that a family of 6 to 8 members typically cooking three hours per day would need about two pounds of oilseed fuelcake daily. About 10 to 16 high-yielding oilseed trees, such as

Calophyllum inophyllum or Sapium sebiferum, might be
sufficient to meet this need. These trees would require
about one-tenth acre in standard orchard configuration.
Alternatively, such trees might be grown around homes or
along fence lines or in other areas not specifically
needed for other crops. In our time trials conducted on
Sapium sebiferum, the Chinese tallow tree, it appears that
about one hour is required for one person to pick and
crush two pounds of seeds. This is perhaps a lighter task
than the wood-collecting chore typically performed by many
rural families in fuelwood-scarce developing nations, in
which long and frequent walks to trees are required.

Village-Level Processing
of Oilseeds for Household Fuels

Background

Although some oilseeds may only need crushing to
produce a crude yet serviceable fuelcake, most will
probably require additional processing. Partial extraction of the oil may be needed to reduce oil percentage to
the optimum level for production of fuelcakes. Alternatively, crushed seeds might be blended with hulls or
other woody materials before being molded into suitable
shapes. Extracted oils will need to be refined before
they can be used as lighting fuels. Although simple, most
of these processing steps probably require expertise and
capital for equipment beyond the resources of most rural
people. Therefore, we believe that production of household oilseed fuels is more likely to be carried out by
small-scale village industry, rather than by the consumer.

Oilseeds have traditionally been processed in village
mills for edible and industrial oils. Many of these well-
known processing methods are relevant to oilseeds for
household fuels. Processing will comprise four basic
steps, essentially: (a) seed storage and preparation, (b)
oil extraction, (c) oil refining and storage, and (d)
blending and molding of fuelcakes.

Seed Storage and Preparation

If incorrectly stored, oilseeds can be plagued by
many post-harvest loss problems. First, seeds must be
properly dried. Wet seeds can be damaged or destroyed by
molds, fungi, and the development of rancidity during
storage, and excess moisture may make processing difficult.
Acceptable moisture levels may be as low as 7 percent by
weight (17). After drying, seeds may be cleaned by winnow-

ing and sieving; specific methods are known (18). Unfortunately, such devices are often inefficient, therefore, Mathieu (4) has developed and tested a new cleaning device which seems to offer advantages over previous approaches.

Some seeds may need to have hulls stripped off by a mechanical decorticator. With the introduction of new oilseeds, new decorticators may need to be designed or old ones adapted. A number of village-level decorticators appear to be suitable or adaptable (19,20).

Good storage facilities are mandatory to curtail losses from deterioration and from pests. A related problem is the danger of spontaneous combustion arising from heat released during deterioration. Simple, low-cost and effective seed storage facilities have been described in the appropriate technology literature (21,22).

Oil Extraction

Six main categories of technologies for the extraction of oil from oilseeds might be considered for village industry: rendering, lever pressing, wedge pressing, screw pressing, hydraulic pressing, and small expellers. All rely on heat or mechanical force, or both. We have not included solvent extraction because of its skill and capital requirements, and inherent safety hazard.

In rendering, oil is extracted by heat without mechanical force. Yields are typically low, but the process is simple. If one needs to partially remove oil to lower the oil content into the 35 to 45 percent range for fuelcake preparation, rendering might be satisfactory. A variety of rendering techniques has been described in the literature (17,23,24).

Simple lever presses (25) generally give yields that are somewhat better than from rendering. Even greater force and higher yields can be attained if the lever is applied through the use of a mortar and pestle driven by an animal or a motor, as in the case of the Asian ghani (3,26,27,28). Ghanis of this type have been widely used in villages throughout the Middle East and the Indian subcontinent, but are declining in numbers as solvent extraction technology has been introduced in central locations for large-scale processing. Village ghanis that remain may depend heavily on uncompensated family labor. The main technical disadvantage of the ghani is that substantial amounts of free fatty acids (FFA) may be

formed (29) and this may not be desirable in oils to be
used for lighting.

Wedge presses, among the oldest methods of seed oil
extraction, can be as simple as the application of gradual
pressure on sacks of cooked crushed seeds by wedges driven
between a plate that bears on the sacks, and an outside
stationary plate. The length of time required to press is
longer than in most modern methods, and wedges and bearing
plates need frequent replacement. Local materials of
construction can often be used, and manufacture would be
simple (30,31).

Screw presses of the traditional type are
among the oldest of oilseed pressing devices. Although
the metal parts involve casting and machining, tolerances
are not demanding and manufacture within developing countries is probably feasible. Maximum pressure is limited
by the strength of the screw, so screw presses should
probably be limited to seeds that do not require high
crushing pressures. Numerous examples of screw presses in
developing nations have been described (18,31,32,33,34).

Most hydraulic presses are probably too large and
expensive for village industry. Small ones do exist (35)
and offer rapid processing at moderate cost, perhaps of
interest for applications in large towns.

Expellers utilize a motor-driven auger to provide
continuous, rapid processing at high pressures. Fric-
tional heat is generated, aiding the extraction process.
Free fatty acid (FFA) formation is low. Disadvantages
include high initial cost and substantial maintenance
requirements. Low-pressure rather than high-pressure
expellers (19) may be more attractive for applications in
developing nations.

Oil Refining and Storage

Crude seed oils need to be refined to prevent excessive deterioration during storage caused by high percentages of FFA, and to ensure proper burning characteristics.
FFA contents over about 5 percent tend to lead to charring
of the wick and production of smokey flames. Other im-
purities such as albuminous materials and particles of
dirt and trash in the oil can clog the wick, and will need
to be removed by filtration (36). FFA can be removed by
contacting the oil with an alkali solution, and allowing
the soap that is formed to settle out. Washing of the
alkali-treated oil and warming to drive off water will then

be required. The oil should then be stored in tins
filled to the top to minimize contact with air and moisture.

Blending and Molding of Fuelcakes

Equipment for these processes will have to be adapted
from other fields or designed as new equipment. There are
few precedents to follow, and development work is needed.
However, difficulties are not expected.

Conclusions

Neither seed oils nor oilseeds presently find significant markets for use as illuminants or cooking fuels.
Most of the lamps in current use have been designed for
kerosene and are, therefore, technically unsuitable for
use with seed oils. New, simple lamps that can utilize
seed oils are probably a feasible design objective.

Whole oilseeds as substitutes for wood or charcoal
are generally unsatisfactory because combustion is too
rapid and the flame is smokey. Preliminary experiments
show that crushed whole oilseeds, tightly packed in the
form of cakes, may be a much more feasible approach but
more experimentation is needed to moderate the rate of
combustion and to decrease smokiness in cookstove service.

Widespread adoption of oilseeds as lighting and
cooking fuels in developing countries might lead to the
generation of new small-scale industries and employment in
rural areas, to process seeds and prepare illuminating
oils and solid fuelcakes from them. A variety of seed
processing techniques is available and probably technically and economically feasible for village application.

The processing of oilseeds need not require high
efficiencies of oil extraction, because the remaining seed
meal must contain a high percentage of oil for fuelcake
formation. Therefore, it may be possible to revive
village oil extraction industries now dying out due to
competition by efficient, centralized and capital-intensive solvent extraction plants.

Acknowledgment

The authors are grateful for the cooperation of the
University of Houston for the use of facilities at the
University's Coastal Center, where the laboratory studies
on wicks and fuelcakes were carried out. Dr. H. William

Scheld provided encouragement and guidance during that phase of the work.

References and Notes

1. The heating value of charcoal (Ref. 5, page 27) is about 12,800 Btu/lb. Chinese tallow tree seeds with nearly 50% fat plus oil measure 12,000 $\pm$ 600 Btu/lb by bomb calorimeter (reference 29 of Chapter 6). Kerosene has a heating value of about 18,000 Btu/lb (G.D. Hobson, ed., Modern Petroleum Technology, 4th ed., Applied Science Publishers, Ltd., Essex, England, 1973). Seed oil heating values are about 17,000 Btu/lb (E.H. Pryde, "Vegetable Oil vs. Diesel Fuel: What Constitutes a Good Vegetable Oil Fuel?", Alcohol and Vegetable Oil as Alternative Fuels, Proc. of Regional Workshops, Northern Agricultural Energy Ctr, USDA, Peoria, IL, (1981) 302.

2. Anonymous, "Potential of Minor Oilseeds in India", The Oils and Oilseeds Journal, (January-March, 1979) 13-14.

3. Anonymous, Oils and Fats in Pakistan, Pakistan Council of Scientific and Industrial Research, Publication Branch, Karachi-3, (1966).

4. S.L. Mathieu, Potential Utilization of Oilseeds for Household Energy at the Village Level, M.S. Thesis, Department of Technology and Human Affairs, Washington University in St. Louis, (May, 1982).

5. D.E. Earl, Forest Energy and Economic Development, Clarendon Press, Oxford, (1975).

6. L. Thwing, Flickering Flames: A History of Domestic Lighting Through the Ages, Charles Tuttle Co., Rutland, VA, (1966) 43-79.

7. F.W. Robins, The Story of the Lamp and the Candle, Oxford University Press, NY, (1939).

8. E. Cecelski, et al., Household Energy and the Poor in the Third World, Research Paper R-15, Resources for the Future, Washington, D.C., (1979).

9. C.M. Duffus and J.C. Slaughter, Seeds and their Uses, John Wiley & Sons, New York, NY, (1980).

10. L. Bell, The Art of Illumination, McGraw Publishing Co., New York, NY, (1902), 56-67.

11. A. Elliott, "The Illuminating Value of Petroleum Oils", Transactions of the Illuminating Engineering Society, Vol. III, (1908) 434-448.

12. Anonymous, "An Incandescent oil-Lamp for School Lighting", The Illuminating Engineer, Vol. IV, (1911) 249.

13. F.E. Cady and H.B. Dates, Illuminating Engineering, John Wiley & Sons, Inc., London, (1925) 40-73.

14. C.P. Steinmetz, Radiation, Light and Illumination, McGraw-Hill Book Company, New York, NY, (1909) 128-137.

15. J.D. Dywer, "Flora of Panama", Annals of the Missouri Botanical Garden, Vol. LII, No. 1, (February 1965) 448.

16. E.W. Schmidt, S.L. Mathieu and E.B. Shultz Jr., "Comparison of Oilseed Fuels with Conventional Fuels in Simple Cookstoves", Paper presented at the 5th National Conference on the Third World, University of Nebraska at Omaha, Omaha, NB, Oct. 27-30, 1982.

17. D. Swern, ed., Bailey's Industrial Oil and Fat Products, Vol.I, 4th ed., John Wiley & Sons, New York, NY, (1979).

18. S.D. Vidyarthi, Production of Vegetable Oils, Jaideva Brothers, Baroda, India, (1951).

19. J. Boyd, Tools for Agriculture: A Buyer's Guide to Low Cost Agricultural Implements, Intermediate Technology Publications, Ltd., London, (1976).

20. M. Carr, Appropriate Technology for African Women, United Nations, (1978).

21. C. Lindblad and L. Druben, Small Farm Grain Storage, Vol. I, Preparing Grain for Storage, VITA, Mt. Rainier, MD, 20822, (1980).

22. Community Development Trust Fund of Tanzania, Appropriate Technology for Grain Storage in Tanzanian Villages, The Advocate Press, New Haven, CT, (1977).

23. A. Martin, The Oil Palm Economy of the Ibibio Farmer, Ibadan University Press, Nigeria, (1956) 11-15.

24. H.K. Dean, Utilization of Fats, Chemical Publishing Co., NY, (1938) 141-157.

25. J.H. Boatwright, How to Get Waterproofing Substances from Plants, Volunteers in Technical Assistance (VITA), Mt. Rainier, MD, (1977).

26. P.F. Knowles, "Processing Seeds for Oil in the Towns and Villages of Turkey, India, and Egypt", Economic Botany, Vol. 21, (April-June 1967) 156-162.

27. S. Rao, A Search for An Appropriate Technology for Village Oil Industry, Appropriate Technology Development Association, Lucknow, India, (1980).

28. S. Chandrasekaran and K.T. Achaya, "Profile of Indian Vegetable Oil Industry, Part 1, Production System", Economic and Political Weekly, (Feb. 23, 1980) 441-45.

29. S.D. Rao et al., "The Composition of Oil-cakes Produced by Different Crushing Equipment in India", A summary published by the Indian Central Oilseed Committee, Proceedings of the First Conference of Oilseed Research Workers in India, (December 15-16, 1958).

30. J.H. Boatwright, "A Wedge Press for Oil Extraction", Appropriate Technology 6, No. 2, (August 1979) 24-25.

31. C.S. Fan, "The Oil Pressing Industry in Hopei Province", Yenching Series on Chinese Industry and Trade No. 3, (1934).

32. P.R. Hale and B.D. Williams, ed., Lik Lik Buk, A Rural Development Handbook Catalogue For Papua New Guinea, Wirui Press, Lae, Papua New Guinea, (1978).

33. G. Toffin, "La Presse a'Huile Néwar de la Vallée de Kathmandau (Népal); Analyse Technologique et Socio-économique," J. d'Ag. Tropicale et de Bot. App. 23 (July 1976) 183-204.

34. S. Corbett, "A New Peanut Oil-Press Machine Fails to Improve Older Methods", VITA News, (April 1981) 4.

35. C.W.S. Hartley, The Oil Palm, 2nd ed., Longmans, London, (1977).

36. J. Lewkowitsch, Chemical Technology and Analysis of Oils, Fats, and Waxes, Vol. II, MacMillan and Co., London, (1915) 1-21.

Eugene B. Shultz, Jr., Robert P. Morgan

Concluding Remarks

The previous chapters have illustrated the remarkable
diversity inherent in the field of oilseeds -- diversity in
habitat, growth habit, types of seed oils and their chemical structures, types of uses, in opportunities for economic
development and in concomitant challenges to wise use of
land and other resources. Thousands of plants, cultivated
and wild, bear enough of some type of oil in their seeds to
warrant possible consideration as economic crops, including
annuals, perennials and trees. Many grow successfully on
arid, hilly, waterlogged or saline soils, and others require
good soil. Virtually every biome of the world is represented. Although much is known about a few oilseed plants
and trees, most have received little attention, and many are
being eradicated by development activities, especially in
the tropics.

The diversity apparent in the study of oilseeds extends
to the molecular level, with an astonishing variety of fatty
acids available from the many triglycerides present in seed
oils. Short and long-chain fatty acids containing a wide
variety of functional groups are found with conjugated and
nonconjugated systems. Whereas much is written about the
relatively simple triglycerides of commerce used mainly as
food oils, the molecular intricacies of the novel and
inedible triglycerides should be made more widely known,
because of the numerous industrial chemical possibilities
presented by fatty acids that include epoxy, hydroxy,
allenic, acetylenic, ethylenic, conjugated and other functional groups and combinations of groups. Many of the
possible chemical products are economically important and
are enjoying substantial growth in demand, including
specialty polymers, synthetic lubricants, plasticizers,
surfactants and antimicrobial agents. We also have the
remarkable possibility that some seed oils may become

important agricultural diesel fuel extenders in some countries, much as Rudolf Diesel had originally conceived.

If oilseeds are to be developed further as chemical and
fuel resources, to complement their already substantial
contribution to world food and feed supplies, many problems
must be solved. In key technical areas the literature is
still incomplete, and research and development is deserving
of far greater support. Examples of needed R&D activities
include oilseed agronomy, especially on marginal lands;
diesel utilization in both direct and indirect injection
engines; development of new industrial products from fatty
acids; and studies of less well-known oilseed species,
especially those with promise as renewable resources of
industrial products of high unit value. Concurrent with
such technical and economic investigations should be a
variety of studies on the sociocultural and management
aspects of oilseed innovations, concerned with evaluation of
the associated risks, benefits and disbenefits, as well as
identification of potentially-significant government inter-
ventions that might enhance or impede innovative development.
Of particular importance is the potential negative impact of
oilseed utilization for fuels and chemicals on the price
and availability of food and animal feed in both the U.S.
as well as developing countries.

When a new field of scientific and technical inquiry
bursts onto the scene, there is often an initial feeling of
euphoria as the many opportunities for new products and
processes become apparent. However, the hard work of sort-
ing out the feasible and practical from the unachievable
then begins. Various authors in this book indicate that
many obstacles must be overcome if oilseeds -- and especially
novel oilseeds not now cultivated commercially -- are ever
to become substantial sources of energy or to increase their
share of the market as chemical feedstocks. For one thing,
overall resource estimates cannot be made with any confidence
because many of the species under consideration have never
been cultivated. For another, the economics of oilseed
utilization is in a very rudimentary stage and differing
conclusions can be reached about specific applications and
ventures, as is indicated by the varying viewpoints of the
authors of this volume.

As this book went to press, concerns about petroleum
shortages and prices have eased somewhat and immediate pressure for considering alternatives to petroleum seems less.
Research and development on renewable energy sources has
been dramatically cut back by the Reagan administration.
But if the time for renewable fuels is not quite here, it

may not be far away. Seed oils are already significant
sources of chemicals and likely to grow in importance, as
Pryde and others have noted. In the case of seed oil fuels,
the economics may not yet be favorable, and alternative
land uses, existing cropping practices, entrenched interests,
and the lack of public policy support also constitute major
obstacles to change.

Further research and development on oilseeds for fuels
and chemicals is needed. It is our hope that this need will
be clearly identified as worthy of both public and private
sector interest and support. In an age of finite, nonrenewable fossil fuels, of continued uncertainty about the
economics and safety of nuclear power and of immense expenditures on military-related research and development, our
biological and botanical heritage in general, and oilseeds
in particular deserve much more attention than they have
managed to receive to date.

Index